Romy Jaster
Agents' Abilities

Philosophical Analysis

Edited by
Katherine Dormandy, Rafael Hüntelmann,
Christian Kanzian, Uwe Meixner, Richard Schantz
and Erwin Tegtmeier

Volume 83

Romy Jaster

Agents' Abilities

DE GRUYTER

ISBN 978-3-11-077750-5
e-ISBN (PDF) 978-3-11-065046-4
e-ISBN (EPUB) 978-3-11-064865-2
ISSN 2627-227X
DOI https://doi.org/10.1515/9783110650464

This work is licensed under the Creative Commons Attribution-NonCommercial- NoDerivatives 4.0 License. For details go to http://creativecommons.org/licenses/by-nc-nd/4.0/.
The Open Access book is available at www.degruyter.com

Library of Congress Control Number: 2020932563

Bibliographic information published by the Deutsche Nationalbibliothek
The Deutsche Nationalbibliothek lists this publication in the Deutsche Nationalbibliografie; detailed bibliographic data are available on the Internet at http://dnb.dnb.de

© 2021 Walter de Gruyter GmbH, Berlin/Boston
This volume is text- and page-identical with the hardback published in 2020.
Printing and binding: CPI books GmbH, Leck

www.degruyter.com

There is no such thing as a 'self-made' [wo]man. We are made up of thousands of others. Everyone who has ever done a kind deed for us, or spoken one word of encouragement to us, has entered into the make-up of our character and of our thoughts, as well as our success.
(George Matthew Adams)

Acknowledgements

The manuscript for this book was finished a long time before the book eventually came out. For carrying me through the dark passage in between I want to thank the Gorgonzola Club; my dearest friends Sanja Dembić, Philip Faigle, Simon Gaus, Kathrin Hasenburger, Jan Hempelmann, and Christian Kobsda. You were an army.

I also want to thank Simon Gaus and Max Seeger for tirelessly discussing the arguments in this book with me and for joining me in thinking more about abilities than any sane person should.

Thanks to Barbara Vetter, whose brain I would sometimes like to borrow; to Christian Nimtz, who is thankfully never content before things are *exactly* right; to Geert Keil, who has been an extraordinarily benevolent and thorough critic of the success view; and to everyone who has given me any kind of feedback throughout the last few years.

This book is for my mom and for the djuns.

Contents

Introduction —— 1

1 Methodology —— 16
1.1 Adequacy condition 1: Extensional adequacy —— 17
1.2 Adequacy condition 2: General and specific abilities —— 19
1.3 Adequacy condition 3: Degrees and context sensitivity —— 23
1.4 Adequacy condition 4: Agentive vs. non-agentive abilities —— 28
1.5 Explanatory challenge 1: Abilities and counterfactuals —— 30
1.6 Explanatory challenge 2: Abilities and dispositions —— 32
1.7 Explanatory challenge 3: Abilities and "can" statements —— 34
1.8 Upshot —— 36

2 The simple conditional analysis —— 38
2.1 Modal semantics: Counterfactuals —— 40
2.2 Extensional inadequacy I: Masks —— 44
2.3 Extensional inadequacy II: Impeded intentions —— 49
2.4 The problem with general and specific abilities —— 52
2.5 The problem with degrees and context sensitivity —— 55
2.6 The problem with non-agentive abilities —— 58
2.7 Upshot —— 60

3 Possibilism —— 63
3.1 Modal semantics: Restricted possibility —— 64
3.2 Possibilism: The details —— 68
3.3 Upsides & downsides I – General vs. specific abilities & masks —— 72
3.4 Upsides and downsides II – Impeded intentions, agentive vs. non-agentive abilities, and other possibilities —— 77
3.5 The problem with degrees and context sensitivity —— 81
3.6 A formal problem —— 84
3.7 Upshot —— 88

4 The success view I – Agentive abilities —— 92
4.1 The general framework —— 94
4.2 The proper motivational state —— 100
4.3 An account of degrees and context sensitivity —— 103
4.4 Impeded intentions and the existential requirement —— 108
4.5 An account of general and specific abilities —— 112
4.5.1 General abilities —— 115

4.5.2	Specific abilities —— 116	
4.5.3	Dependence —— 121	
4.5.4	Success and failure —— 124	
4.5.5	Degrees of specificity —— 126	
4.5.6	Hypothetical circumstances —— 128	
4.6	Masks —— 135	
4.7	Formal considerations revisited —— 137	
4.8	Finks and other unintentional abilities —— 140	
4.9	Impairments by ignorance —— 146	
4.10	Upshot —— 150	
5	**The success view II – Non-agentive abilities —— 154**	
5.1	The success view for non-agentive abilities —— 155	
5.2	Degrees and the distinction between general and specific abilities —— 157	
5.3	The broader scheme: the success view of abilities tout court —— 159	
5.4	Abilities and dispositions —— 162	
5.5	The normativity of abilities —— 168	
5.6	The existential requirement revisited —— 170	
5.7	Proportions among infinite sets —— 173	
5.8	A brief reflection on the explanatory challenges —— 176	
5.9	Upshot —— 177	
6	**The success view situated —— 180**	
6.1	Success 1.0 —— 185	
6.2	The sophisticated conditional analysis —— 188	
6.3	Possibilism+ —— 199	
6.4	An option-based account —— 205	
6.5	Upshot —— 212	
7	**The success view applied – Two rough sketches —— 214**	
7.1	A paradox about time travel revisited —— 215	
7.2	Alternate possibilities contextualism about freedom —— 218	

Resumé and an open question —— 225

References —— 229

Index of Names —— 234

Index of subjects —— 236

Introduction

I cannot dance Tango, but some people can. My partner can play intricate songs on the guitar. My dog can locate a hidden toy by sniffing around the apartment. Some people can do a handstand, but very few can do it while drunk.

The statements I just made talk about agents' abilities. They state of some agent (or group of agents) that they have or lack the ability to perform a certain action. In a way, ability statements are not particularly extraordinary; we ascribe abilities to agents all the time, we have no trouble understanding ability ascriptions, and some of those ascriptions, like the exemplary ones above, state plain and obvious truths.

Trouble looms once we put our philosopher's hat on. For as philosophers, of course, we will want to know what makes ascriptions of ability true. What is it for some agent to have an ability? What, in other words, are the truth conditions for statements like the ones we started out with? Here, the answer is far from obvious.

What is clear is that the answer will have to appeal to modality in some way or other. That is because apparently an agent need not *actually* perform a certain action in order to have the ability to perform it. Yet, performing the action surely has to be open to the agent in some sense – it has to be somehow possible for her to do what she is able to do. In this sense, abilities are *modal properties:* they have something to do with the agent's behavior in possible scenarios.

At this point, we are already knee-deep in the swamp of philosophical intricacies. For, as a closer look quickly reveals, the modal truth conditions for ability statements are actually hard to spell out. Some have tried to elucidate abilities in terms of some kind of restricted possibility. Others have appealed to the modality of counterfactuals. I say: both attempts are bound to fail. The modality of abilities is more complex than that. The quest for the right truth conditions will therefore take us beyond the common schemes of possibilities and counterfactuals very quickly.

Why should we bother with abilities to begin with? One answer is: that's just what we philosophers do. We try to wrap our head around interesting phenomena. We want to understand what knowledge is, what rationality is, what truth is, what meaning is. Surely, then, one natural question to ask is: what are abilities?

But there are further good reasons to yearn for a better understanding of abilities. First, and obviously, abilities play a fundamental role in a wide range of philosophical theories. Concepts (Millikan 2000, ch. 4; Evans 1982: 104), qualia (Lewis 1990), knowledge (Greco 2009, 2010: 3; Sosa 2015: part II), actions (Mayr 2011, ch. 7–9), conceivability (Yablo 1993; Menzies 1998) – all of these phenomena and many more have been crucially linked to or even analyzed in terms of the notion of abilities. A better grasp of what it is for an agent to have an ability will yield a better understanding of all of those views insofar as a core notion figuring in the views will be more transparent.

A second motivation for a closer look at abilities is that abilities are closely related to dispositions, which have attracted a lot of interest recently. Both dispositions and abilities are modal properties in the sense that they have to do with the subject's behavior in possible situations. For abilities, we have already noted this fact: to have an ability to φ, one need not actually φ. Dispositions exhibit an analogous modal structure. Sugar has the disposition to dissolve in water, say, even if it is not actually submerged in water and thus does not actually dissolve. This similarity and others (→ 1.6) have led some philosophers to think that abilities simply *are* dispositions of sorts (i.e. Ryle 1949; Moore 1912; Fara 2008; Vihvelin 2004, 2013; Smith 2003). This may or may not be true; in any case, abilities should be of interest to anyone interested in dispositions, because it may well turn out that the two phenomena call for an analysis along roughly the same lines.

Abilities should also attract the interest of anyone interested in modal semantics.[1] One (and as we will see, the most paradigmatic) way abilities can be ascribed is via the modal auxiliary "can". "Ela can cite the poem by heart"; "Peter can pick locks in no time"; "I cannot do a summersault here and now". All of those statements are statements about an agent's abilities. And all of them contain the modal auxiliary "can". This is interesting, because there is a strong tendency in linguistics and formal semantics to try and provide a unified semantics of statements involving that modal auxiliary. Ability statements therefore provide an important test case for such a semantics. Whatever form that semantics takes, it had better cover "can" statements ascribing ability. Thinking about the truth conditions of ability statements is therefore mandatory for anyone interested in the semantics of "can" more generally.

Finally, abilities should be of interest to anyone interested in the free will problem. One of the great impasses in the literature on freedom, and perhaps one of the most tightly woven knots in philosophy as a whole, traces back to a controversy about what it is for an agent to have an ability to act otherwise. The problem is well-taken. Even post Frankfurt (1969), one cannot help but feel that free will requires the ability to do otherwise.[2] And that ability seems ob-

---

[1] For the standard view cf. Kratzer (1977, 1981).
[2] In this seminal paper, Frankfurt famously argued that responsibility does not require alternate possibilities. This finding is often extended to the conclusion that freedom, in the sense relevant

viously incompatible with determinism. No one could have done otherwise, after all, if every single one of everyone's acts is determined to happen by the state of the world and the laws of nature. Hence, the argument goes, free will is incompatible with determinism. We either lack free will or determinism has to be false (locus classicus: van Inwagen 1983). This is one side of the debate; the side of the incompatibilist.

The other side is that of the compatibilist. Compatibilists reject the line of reasoning employed by the incompatibilist, and not just because it leads to strange consequences. Instead, one typical compatibilist argument has it that the ability to act otherwise *is* in fact compatible with determinism. How so? Well, here compatibilists diverge. The classic idea, and one that has become something like the default view in certain areas of philosophy, is that to have an ability, and *a forteriori* the ability to act otherwise, it has to be true that one would φ, if one were properly motivated to φ (locus classicus: Moore 1912, ch.1). This view, which is known as "the simple conditional analysis of abilities", will be discussed extensively in chapter 2. A more recent idea is that to have an ability to φ, one has to have a disposition (or set of dispositions) of sorts, a view which has recently regained new popularity under the heading "new dispositionalism" (Fara 2008; Vihvelin 2004, 2013; Smith 2003) (→ 5.4, 6).

Whether one of those views is correct or not need not concern us at the moment. What is important is that both of the conditions just mentioned seem to be compatible with determinism. The counterfactual can be true, even if the agent is determined not to φ; and the disposition to φ when one is properly motivated to φ can be had, even if it is determined that it does not manifest itself. This offers a straightforward argument against the incompatibilist: contrary to appearances, the ability to act otherwise *is* in fact compatible with determinism, and hence the fact that freedom requires the ability to act otherwise does not rule out the existence of free will in a deterministic world.

Who is right? This is one of the core questions in the free will debate. But to answer it, one will obviously need an account of abilities generally. How, after all, could one possibly decide whether or not the ability to act otherwise is compatible with determinism, if one does not have a firm grip on what it is for an agent to have an ability to do anything whatsoever? This, then, is the fourth line of motivation for a closer look at abilities. Coming to terms with free will requires coming to terms with abilities first.

---

for responsibility, does not require alternate possibilities. Whether or not Frankfurt's arguments validate either of the two conclusions is a matter of thriving controversy. For some important contributions, cf. Widerker & McKenna (2003).

In view of all that, it is rather surprising that in-depth discussions of abilities themselves are sparse in the literature. Most remarks on abilities are rather cursory and scattered across various more or less unconnected debates; the few more elaborate accounts in the literature are still rather sketchy; and even the two most influential views of abilities are strangely insulated from one another in that they are situated in two different parts of the philosophical landscape and have been developed against the background of two different systematic endeavors.

On the one hand, there is the before-mentioned simple conditional analysis, according to which having an ability is for a certain counterfactual to be true.³ On the other hand, there is what I will call "possibilism", a view according to which having an ability is for a certain restricted possibility to obtain. Both views are dominant in the literature in that both enjoy the status of a default view in certain areas of philosophy. Interestingly, though, there is little exchange between the adherents of the two views and, as far as I can see, there is virtually no assessment of both views at once.⁴

That is odd, to say the least, and it is unfortunate, given that both of the two views voice important insights about abilities which elude their respective counterparts completely. Odd as the situation is, though, it is not mysterious. There is an obvious explanation for the striking division in the literature. And that explanation, I take it, is that the two most dominant views of abilities have been developed with an eye towards very different systematic goals.

The simple conditional analysis has its origins in the free will debate, and more specifically in the context of the compatibility problem of free will and determinism. Both Moore (1912, ch. 7), and before him Hume (1748, 8.23/95), famously argued that determinism is compatible with the ability to act otherwise. The reason for that, they thought, is that abilities, and thus the ability to act otherwise, have to be understood in terms of a conditional.

Hume thought that for us to have the power to move, it has to be the case that "if we choose to move, we (…) may" (ibid: 72). Moore gave a counterfactual reading to this condition and argued that for an agent to have an ability to act otherwise (in the sense relevant not just for free will, but also for actions to

---

3 There is also what I will later call "the sophisticated conditional analysis", according to which abilities have to be analyzed in terms of a multitude of highly complex counterfactual conditionals (Manley & Wasserman 2008; Sosa 2015: 98). While the sophisticated conditional analysis clearly stands in the tradition of the simple conditional analysis, the view is so recent in the literature and differs so considerably from its historic source that I will discuss it separately later on (→ 6.2).

4 An exception is (Lehrer 1976).

be voluntary) is for it to be true that the agent would have acted otherwise, had she so chosen (Moore 1912, ch. 1). Moore's work, in particular, has had major impact in the philosophy of action and free will and was nothing short of groundbreaking for people's thinking about these issues. It is therefore unsurprising that the conditional analysis has become the default point of reference for most philosophers with a background in one of those areas.[5]

The other systematic goal that has guided people's thinking about abilities is a very different one: it is the task of coming up with a viable modal logic and modal semantics. Here, the key project is to understand the logical and semantic workings of modal auxiliaries like "can", "must", "may" and so forth. Since abilities are paradigmatically ascribed via "can" statements, one subproject of this overarching endeavor is to embed the logic and the semantics of ability statements into whatever turns out to be the most plausible logic and semantics, respectively, of the modal auxiliaries quite generally.

It is in this connection that possibilism has flourished. The starting assumption both in semantics and logics has been that "can" expresses possibility, and "must" expresses necessity. Logicians like von Wright (1951) and linguists like Kratzer (1977, 1981) have built whole systems from there. And since ability ascriptions have been taken to fall squarely on the possibility side of that dichotomy, possibilism has become the default point of reference for most people with a background in logics, semantics, or, generally, the more formally inclined fields of philosophy.[6]

This, I take it, is the situation. There are two views, each of which is plausible enough to have become a default view in its respective field of application. There is little to no exchange between adherents of the two views. And there is no way for the two views both to be true at once. A systematic, careful and unprepossessed in-depth examination of the subject matter is thus far lacking.

The book at hand sets out to change this. In what follows, I provide a systematic and comprehensive account of agents' abilities, or, more specifically, of what it is for some agent to have the ability to engage in some behavior. Paradigmatically, the behavior which an agent has an ability to perform will be an

---

[5] That is not to say that *everyone* in this field endorses the simple conditional analysis or a view in that tradition. See for instance Horgan (1979) and Lehrer (1976) for a possibilist treatment of the ability to act otherwise.

[6] Again, that is not to say that possibilism has been *limited* to the formal treatment of ability statements. The view has been applied to the problem of time travel (Lewis 1976), and also, as noted before, to the problem of free will (Horgan 1979; Lehrer 1976). Its origins, however, lie in the formal treatment of "can" statements and it is in the formally oriented fields that possibilism has the status of a default view.

action, in which case I will speak of *agentive* abilities. In other cases, the ability is one to be engaged in behaviors that are not actions, such as smelling, digesting, or unintentionally reading street signs,[7] in which case I will speak of *non-agentive* abilities. The view I will develop in this book accounts for both agentive and non-agentive abilities alike.

As we'll see, that view will be inspired by core ideas of both of the two views of abilities that have dominated the literature so far. But it will be superior to both of those views in that it will be apt to meet the adequacy conditions for a comprehensive view of abilities and solve the problems that arise with each of the other two views in particular.

The core idea underlying the account I put forward is that abilities are a matter of success across a sufficient proportion of modal space. For this reason, I will dub the view "the success view of ability", or simply "the success view" – it is to be hoped that *nomen est omen* in this case.

Before walking you through the line of argument that is going to be in the book, let me say a few words about what I am trying to accomplish. What I am after, throughout the book is an analysis of what it is for an agent to have an ability to φ. And by an "analysis" I mean informative and illuminating truth conditions of canonical ascriptions of abilities. What I will try to provide, then, is a set of necessary and sufficient conditions for an agent to have some ability to φ. Or, put differently again, I will try to fill the blank in the statement "An agent has an ability to φ if and only if _____".

Note that in discussing other accounts of abilities, I will treat those views as answers to the question of how to fill that blank as well. I will, in other words, discuss the existing views of abilities as analyses in the specified sense. I will not consider whether they are fruitful contributions to any other philosophical project, such as stating merely a necessary condition for having an ability or, less ambitious still, highlighting a merely typical but systematically interesting feature of abilities.

Perhaps you are one of the philosophers who sense methodological pitfalls when it comes to the project of coming up with an analysis. The project may seem hopeless to you; philosophy does not provide us with many examples of successful analyses, after all. Or you may be skeptical about intuitions and the conceptual competence of philosophers, which makes you worry about the solidity of the foundations on which analyses are often built. Both worries are quite common these days. So let me say a few sentences about each.

---

[7] The example of the ability to read street signs unintentionally is brought up by Löwenstein (ms.). I'll make frequent use of this very fruitful example in the course of this book.

As to the pessimism about the prospects of coming up with a successful analysis of anything at all, I am inclined to respond that even though attempts for comprehensive analyses may regularly fail, engaging in the quest for analyses nevertheless brings progress. If there is a fact of the matter about whether some x is F, then it should, in principle, be stateable which fact of the matter that is. Things are terribly complex, of course, which is why stating the fact of the matter that has to obtain for x to be F is typically extremely difficult. But trying to state it is nevertheless worthwhile, because it forces the mind to be maximally strict. And that results in lots of insights that would otherwise have escaped us.

Suppose no one had tried to analyze knowledge. Instead, philosophers would have contented themselves with stating something along the lines of: "knowledge has a lot to do with belief, justification, and truth". First, of course, it is not very likely that anyone would have taken the time to formulate this mundane insight at all. But if they had, then there would not have been any reason to object. Gettier would not have had any reason to think hard of cases in which something is going on which cannot be captured in terms of belief, justification, and truth. And maybe, just maybe, it would still not have occurred to people that luck eats knowledge for breakfast. I therefore see just as much value in *trying* to state an analysis as I see in the analysis itself.

Moreover, I take it that there is, in fact, progress being made in stating analyses in philosophy. Take again the example of knowledge. To my mind, at least, it is quite obvious that analyses have become better since Gettier's days. People have understood more about the relationship between knowledge and luck, the role of reliability, and the semantics of knowledge ascriptions, to name but a few dimensions on which understanding has progressed. And even though the analyses that have been built on these insights still admit of counterexamples and are far from uncontroversial, I take it to be clear that they are superior to the pre-Gettier "justified true belief" analysis of knowledge. They are superior, because they are more insightful.

This is where the second worry enters the stage. All good and well, you may think – perhaps an analysis need not be impeccable for it to be insightful. But how on earth are we going to find out about the truth conditions of ability ascriptions, or any philosophically interesting ascriptions at all, for that matter? Is this one of those books in which every argument takes the form "We would not say 'x is F' in this or that highly intricate, bizarre, and outlandish example; hence, the case is not one in which x is F"?

Let me assure you: it is not. As you will see in the next chapter, the methodology I work with is carefully chosen to avoid walking on the thin ice of intuitions about what "we" would say in bizarre cases. There is little need for intuition pumps about word use in this book. For as we'll see in the next chapter, we can formulate quite a few adequacy conditions for a comprehensive view of abilities, which do not depend on shaky linguistic intuitions and which will serve as guide rails for the discussion to follow.

Let me also emphasize what I won't do. First, although I will be working with a possible worlds framework throughout the book, I am explicitly neutral on the metaphysics of modality. I view the possible worlds framework I will be working with as *just that:* a framework. As philosophers, our job is to look at distinctions and patterns that elude the untrained eye and which therefore often cannot be described by using ordinary language. Ordinary language has evolved to draw distinctions and describe patterns that matter in everyday life. It does not provide the means to speak clearly about the subtleties we need to pay attention to in philosophical discourse. One purpose of formal frameworks, I take it, is to provide a way of sharpening our language in such a way that philosophically interesting distinctions can be drawn, and patterns described. When it comes to modal notions, such as that of abilities, a possible worlds framework is simply a very powerful framework to work with.

It is also the framework I know best. By using it, I do not want to commit to any particular view about the underlying metaphysics of modality, nor do I want to commit to any particular metaphysics of possible worlds themselves, nor to any claim to the effect that the metaphysics of modality is one of possible worlds to begin with.

If you yourself have strong metaphysical views, you can take this as an invitation: whether your favorite modal metaphysics is one of possible worlds (Lewis 1986), one of potentialities (Vetter 2015), or of something else entirely – as long as your metaphysics respects the patterns and distinctions I am trying to establish in the course of the book, I invite you to plug in whatever picture of the metaphysics of modality you favor.

I will also abstain from the metaphysical question of whether abilities are *fundamentally* categorical or modal properties; I don't want to commit to any view on whether there can be abilities without a categorical basis. Questions like these were dominant in the literature on dispositions for a long time (Armstrong, Martin & Place 1996; Mackie 1973, 1977; Prior, Pargetter & Jackson 1982). People wanted to know whether having the disposition to break, say, *is* having a particular molecular structure or whether the disposition to break is instead a modal property which, while possessed in virtue of that molecular structure, is nevertheless distinct from it.

It was Lewis's (1997) seminal paper that shifted the focus from this question to the project of coming up with a viable analysis of dispositions. Lewis did not really care whether or not breakability is ultimately realized by a categorical or a modal property – in fact, that question can be answered either way, on his view.

Instead, he was interested in the truth conditions for having a disposition. My interest in abilities is analogous to Lewis's interest in dispositions in this respect. I want to know what the truth conditions for canonical ability ascriptions are; I am not interested in the question whether agents' abilities are fundamentally based in categorical properties or in genuinely modal ones.

It is instructive to compare what I am and am not doing to the two projects a functionalist about mental properties can pursue. Functionalists can be interested in two distinct questions. One question is: what are the truth conditions of canonical ascriptions of being in pain, say? In response to this question, the functionalist proposes that a subject is in pain if and only if the subject is in a state with the causal pattern characteristic of pain states. The question I am pursuing regarding abilities is analogous to this project. I want to know what the truth conditions of canonical ascriptions of ability are. In response to this question, I will propose that an agent has an ability if and only if the agent has a property with the modal pattern characteristic of abilities.

Some functionalists, however, go on to ask a second question. The second question is: which state(s) do(es) actually play the specified causal role? Several sub questions come up: is it a physical state? A brain state? Is it just one or are there many? And so forth. Likewise, I could in principle go on and search for the property that underlies the specified modal pattern of abilities. Again, several sub-questions come up: is it a categorical or a modal property? Is it the same kind of property for every type of ability? And so forth. These are interesting metaphysical questions, no doubt. But I take it to be obvious that they specify a different project from the one specified by the other question I formulated. One is the project of providing truth conditions of canonical ability ascriptions. The other is that of settling the metaphysics of the subject matter. I am after the analysis and neutral on the metaphysics.

Why do I care so much for a viable analysis of ability ascriptions? On the one hand, it seems to me to be the only plausible place to start thinking about abilities. Even if one's primary goal were to find out what kind of properties abilities *are*, metaphysically speaking, one would, I take it, first have to provide a job description for the properties one is hoping to find. And that job description is provided by stating the truth conditions for having the property one is looking for. Only then can one begin to check which kind of property is suited to do the job. Coming up with an analysis of the concept of an ability thus seems like a natural place to start one's philosophical investigations.

On the other hand, we will see that the analysis brings all kinds of interesting insights to light. By the end of the book, we will know quite a bit about the relation between abilities and possibility; between abilities to counterfactuals; between abilities and dispositions; between paradigmatic and less paradigmatic instances of abilities; we will know which role intentions and performances play in having abilities; how abilities relate to success and failure; and a lot more. Formulating truth conditions for ability ascriptions is by no means the only interesting approach to abilities. But it is one very fruitful one and the one I am going to pursue.

I will work towards a viable analysis of abilities on two methodological levels, which will sometimes merge. One is the level of our semantic knowledge about the truth and falsity of particular ability statements. (This is in line with what I said about linguistic intuition pumps earlier, in so far as we will not look at ability statements that are made in complicated or bizarre scenarios; the truth and falsity, respectively, of the statements we will look at will be very obvious.) The other is the level of our knowledge about whether or not some agent has a particular ability. These are two sides of the same coin. If I know that "S can ϕ" is true (and I understand that sentence), then I know that S can ϕ, and *vice versa*. I will therefore switch back and forth between metalinguistic talk about ability statements and object talk about abilities a great deal in the course of the book. If you have a strong inclination towards framing philosophical inquiry in one of those two ways, feel free to translate my elaborations into your favorite jargon whenever necessary.

So much for the preamble. Let me now proceed to give you an outline of the line of argument I'll develop in this book.

Chapter 1 lays the methodological foundation for the whole discussion to follow in that it establishes what I will call *adequacy conditions* for a comprehensive view of abilities. The adequacy conditions are conditions that will have to be met by any theory of abilities in order for the theory to capture its subject matter appropriately; being test conditions for accounts of abilities, they will hence play a crucial methodological role in the critique of existing views as well as the construction of a superior view.

The adequacy conditions themselves, I hope, are uncontroversial. There are four of them. The first is that any comprehensive view of abilities will have to be extensionally adequate; the second is that any comprehensive view of abilities will have to elucidate the distinction between so-called general and specific abilities; the third is that any comprehensive view of abilities will have to account for the fact that abilities come in degrees and that ability statements exhibit a certain kind of context sensitivity that goes along with their gradable nature; the fourth is that any comprehensive view of abilities will have to tell us something interesting about the relation between agentive abilities, i.e. abilities to perform actions, and non-agentive abilities, i.e. abilities to be engaged in behaviors that are not actions.

Once laid out, all of those conditions should be obvious. Yet, actually laying them out brings a lot of progress. Starting from what we want a comprehensive view of abilities to deliver, we will have a common ground on which to argue for and against certain features of such a view – a common ground which has so far been completely amiss in the literature.

Besides outlining these adequacy conditions, chapter 1 also develops three explanatory challenges for an account of abilities – questions, that is, towards which one can expect a comprehensive view of abilities to take a stance. The first is how abilities relate to counterfactuals. The second is how abilities relate to dispositions. The third is if and how ability statements can be embedded into a more general semantics of "can" statements. The role of the explanatory challenges is less central to the project of coming up with a comprehensive view of abilities, insofar as the explanatory challenges only really arise once we have such a view on the table. They do not provide test conditions for a comprehensive view, but questions one will want such a view to answer once it is formulated. Yet, formulating the explanatory challenges will help us get a better grip on some important features of abilities and is therefore called for right at the beginning of the book.

Chapters 2 and 3 turn to discussing the two most prevalent views of abilities on the market, pointing out their appeal, but also pointing out that they fail and why they fail. All this is done against the background of the adequacy conditions from chapter 1.

Chapter 2 deals with the so-called simple conditional analysis, according to which having an ability is a matter of some counterfactual conditional to be true. In its most prominent form, the simple conditional analysis states that an agent has an ability to perform an act φ if and only if the agent would φ if she intended to φ. There are several variations of this account on the market, all of which have in common that they use a counterfactual conditional to postulate a modal tie between an agent's motivational state to φ and the agent's performance of φ-ing.

The simple conditional analysis has strong appeal. Abilities have something to do with success. And success, in turn, has something to do with the tie between an agent's intentions (or motivations in general) and her corresponding performances. It is therefore no wonder that the simple conditional analysis has become something like the default view in certain areas of philosophy.

Nevertheless, I will argue that the simple conditional analysis is fraut with problems. In fact, any version of the view will inevitably run into trouble with all of the four adequacy conditions laid out in chapter 1. As to extensional adequacy, I will argue that the simple conditional analysis runs into trouble with abilities that are masked as well as with cases in which the agent cannot form an intention to φ to begin with. Moreover, the simple conditional analysis lacks the means both to elucidate the distinction between general and specific abilities and to account for degrees of abilities and the corresponding kind of context sensitivity that attaches to ability statements. Finally, the simple conditional analysis fails to capture non-agentive abilities. All this strongly suggests that abilities are not properly accounted for in terms of a single, simple counterfactual conditional.

Chapter 3 yields an equally bad prognosis for the other very prevalent view in the literature on abilities: a view I will refer to as "possibilism". Possibilism is the view that an agent has an ability if and only if it is possible, in a properly restricted sense, for the agent to φ. Contrary to the simple conditional analysis, possibilism does not put any emphasis on the modal tie between motivation and performance. Instead, all the emphasis is put on the possibility of the performance.

Like the simple conditional analysis, possibilism is driven by an appealing thought, though a different one: the idea that abilities are always had in view of certain facts. An agent may be able to play tennis in view of her basic physical constitution, say, but not in view of the fact that her wrist is broken. Starting off from this fundamental insight, it is then argued that having an ability is a matter of compossibility with those varying sets of facts and whether or not we count an agent as having an ability will vary, depending on the sets of facts we are interested in.

As I will argue, possibilism looks somewhat more promising than the conditional analysis on first examination, because it avoids the extensional problems that beset the conditional analysis and has no trouble accounting for non-agentive abilities. Going through the adequacy conditions one by one, however, we will see that possibilism is just as problematic as the conditional analysis. It, too, fails to account for the distinction between general and specific abilities as well as for degrees and the corresponding kind of context sensitivity of ability statements. On top, it runs into a deep-rooted formal objection. Together, these difficulties show clearly that abilities are not properly accounted for in terms of restricted possibilities either.

The conclusion of chapter 2 and 3 is straightforward, then: given a plausible set of adequacy conditions for a comprehensive view of abilities, the two major views of abilities, while both voicing an appealing intuition, seem misled in fundamental ways.

Luckily, we will have learned enough about abilities at this point for a more informed way of thinking about abilities to take shape. In chapters 4–6, I lay out a new view of abilities: the success view. The major intuition driving the success view is that having an ability is a matter of success across a sufficient portion of modal space.

Chapter 4 develops this idea for the paradigm cases of abilities: abilities to perform actions, or *agentive abilities*, as I will call them. As I will argue, having an ability of this kind is for the agent's intentions to be suitably tied to the agent's performances across the portion of modal space that matters. More specifically, I will argue that an agent has an agentive ability to φ if and only if the agent φ's in a sufficient proportion of the relevant possible situations in which the agent intends to φ. Later in the chapter, I will provide a somewhat weaker version of the view, which also accommodates some special cases: on the weaker version of the success view of agentive abilities, an agent has an agentive ability to φ if and only if the agent φ's in a sufficient proportion of the relevant possible situations in which the agent intends to ψ, where φ and ψ need not specify the same action type – as long as φ-ing in response to ψ counts as a success. For simplicity, I will work with the stronger version throughout the largest part of the chapter, though.

As chapter 4 proceeds, it will become clear that the success view has lots of merits. On the one hand, the view combines the core idea driving conditional analyses with the fundamental insight underlying possibilism: the success view does justice both to the idea that abilities are had in view of certain facts and to the thought that agentive abilities are a matter of a modal tie between intentions and performances.

On the other hand, the success view is apt to meet all of the adequacy conditions for agentive abilities. It yields an account of degrees and context sensitivity, it does not run into extensional problems with cases in which the agent cannot form the intention to begin with, it provides a unified account of general and specific abilities, it does not run into extensional problems with masked abilities, and it avoids the formal problem that beset possibilism. In conclusion, we can note that the success view yields a much better understanding of agentive abilities than both the simple conditional analysis and possibilism.

Chapter 4 closes with the discussion of two objections to the success view of agentive abilities, which will demand the modification I pointed out earlier. Looking closely, for an agent to have an agentive ability, it does not have to be the agent's intention *to φ* that is properly modally tied to performances of φing – it may well be some other intention. This will yield a somewhat weaker reading of the success view of agentive abilities than the one I worked with throughout most of the chapter. On that weaker reading, S has an ability to φ if and only if S φs in a sufficient proportion of the relevant possible situations in which S intends to ψ, where φ and ψ need not be identical – neither *de dicto*, nor *de re*. I will offer a way to integrate the original and the weaker version of the success view of agentive abilities in the course of chapter 5, where it will become clear that both versions are instances of the success view of abilities *tout court* (covering the various kinds of agentive abilities as well as non-agentive abilities).

In chapter 5, I will turn to non-agentive abilities. As I will argue, abilities are quite generally a matter of success across a sufficient portion of modal space. In the paradigmatic cases of agentive abilities, that is to say that S has an ability if and only if S φ's in a sufficient proportion of the relevant possible situations in which the agent intends to φ. In the case of non-agentive abilities, a structurally analogous analysis can be provided. On the view I will lay out in chapter 5, S has a non-agentive ability if and only if S φ's in a sufficient proportion of the relevant possible situations in which some S-trigger for φ-ing is present, where an S-trigger is a trigger for φ-ing, in response to which φ-ing is a success.

In the case of agentive abilities, that boils down to the analysis developed in chapter 4, because an action is a success when it follows an intention to perform that action (or another suitably related intention, which is why the success view of abilities tout court covers the weaker version of the success view of agentive abilities as well). In the case of non-agentive abilities, like the ability to digest, hear, or read street signs unintentionally, the triggers in response to which such behaviors are a success will usually be external features of the situation: that the agent ingests food, is subjected to sound, or faces a street sign, say. Structurally, however, there is no difference between agentive and non-agentive abilities. Rather, the success view for agentive abilities turns out to be the paradigmatic instance of an overarching analysis of abilities *tout court*.

Chapter 5 goes on to explore the relationship between abilities and dispositions, reflect on the normative dimension of abilities that enters via the notion of success, recapitulate the responses to the explanatory challenges that have emerged throughout the previous chapters, and respond to a formal objection against the success view. By the end of chapter 5, the success view is fully developed.

In chapter 6, the success view is situated within the contemporary literature. The simple conditional analysis and possibilism are two very influential views in the literature, but they are by no means the only views out there. In fact, there have recently been new suggestions of how to analyze abilities, most of which have emerged against the background of the new dispositionalism, according to which having an ability to φ is a matter of having a (set of) disposition(s) of sorts, and certain virtue-epistemological accounts of knowledge, according to which knowledge of p is a matter of believing that p in virtue of one's intellectual abilities.

In the course of the chapter, we will look at four views, two of which are very much in line with the success view and the other two of which fail for independent reasons. By the end of chapter 6, it should be clear that the success view is a very strong contestant in the literature on abilities.

Chapter 7, the last chapter of the book, gives a quick resumé and offers an outlook of how the success view may actually be put to work. Two realms of application are sketched. One is the grandfather paradox about time travel, where the success view can be used to provide a more informed version of Lewis's (1976) solution to the paradox. The other one is the problem of free will, where I will use the success view to sketch what I am going to call "alternate possibilities contextualism" – a contextualist view of freedom ascriptions which allows for a conciliatory perspective on the ongoing debate between compatiblists and incompatibilists about freedom and determinism.

1 Methodology

Peter Morriss, in a book on power, starts his investigation with what seems to be a platitude:

> Before we can start constructing an account of power we need to know what sort of thing we are dealing with: we must decide just *what* it is that we are trying to analyze. And we must decide, as well, how we go on about deciding *that*. (Morriss 2002: 2)

Why even state that platitude? Well, because, as Morriss points out,

> [m]ost writers (...) pay far too little attention to these preliminary problems, with the result that they go rushing off in the wrong direction, pursuing the wrong quarry. When they eventually catch it, they may claim to have caught the beast they sought; but how do they know, if they didn't know what they were looking for? (ibid.)

What Morriss suggests here is that the debate on power suffers from a severe lack of methodological underpinnings. I am quoting him here, because as far as I can see, the same remarks apply to the literature on abilities. Methodologically, the debate is quite a mess. What is it that we are seeking to account for? How do we test whether an account captures what it is supposed to capture? What is the common ground from which we can start building competing views? These questions, while essential to the project of developing a viable understanding of abilities, do not seem to have played much of a role in the debate thus far. That is unfortunate. In fact, my guess is that the whole debate would have taken a different course had the preliminary issues been settled beforehand. Let us try and do better. Let us do some methodological groundwork before we start.

Perhaps you are inclined to think that any such project is bound to fail and that Morriss's suggestion to specify what we are trying to analyze *before* going about analyzing it is up a blind alley. Isn't this just the paradox of analysis? How are we supposed to state what we are trying to analyze if we don't have the analysis at hand yet? Isn't the dog chasing its own tail here?

It is not. Here is what I think we should do. First of all, let us collect a few things we know about abilities. Obviously, we don't need to have an analysis of abilities for that. We know that dogs wag their tails when they are happy without having an analysis of doghood. In much the same way, there are a few features of abilities that I take to be beyond dispute. Those features need to be identified. This will help to get a better grip on the phenomenon we are seeking to understand and moreover will provide the basis for the entire discussion to follow.

By finding features that are essential for our understanding of abilities, we will get what we can call *adequacy conditions* for theories of abilities – conditions that will have to be met by any theory of abilities in order for the theory to capture its subject matter appropriately. These adequacy conditions serve as a blueprint for the critique of the most prevalent views on the market and for the development of a superior alternative. They provide test conditions along which the prospects of various potential accounts of abilities can be discussed and evaluated.

Apart from those adequacy conditions, I think we can also identify what I will call *explanatory challenges* for a theory of abilities. It is quite obvious upon reflection that abilities relate in interesting ways to certain other phenomena which have themselves been the subject of philosophical theorizing. An obvious challenge for a theory of abilities is to spell out how abilities relate to those other phenomena; in other words: to situate abilities in the broader theoretical landscape. While less essential to the task of formulating a viable view than the adequacy conditions, the role of the explanatory challenges must not be underestimated. Locating a phenomenon's position in a web of other phenomena clearly helps get a more thorough grip on the phenomenon one is seeking to understand.

The adequacy conditions concern features of abilities that have to be captured, then. The explanatory challenges concern the relation of abilities to other phenomena with which they are, for some reason or other, interestingly linked. Formulating a set of each will provide us with a solid foundation from which we can start our investigation.

1.1 Adequacy condition 1: Extensional adequacy

The first thing we can note about abilities is that we know quite a bit about their distribution, so to speak. We know that virtually everyone has *some* abilities, for instance. We know that there are some things pretty much anyone can do, like making sounds. Yet, we also know that it is certainly not the case that anyone can do anything. There are some things no human can do, like flying faster than light. And we also know that we ourselves can do some things, while lacking the ability to do others. I know that I can play the guitar a bit, but cannot play the piano, for example. I have similar knowledge about others. I know that most people I know can swim, say, but that my grandmother cannot. This gives us a first adequacy condition for a comprehensive view of abilities:

> A1. Any comprehensive view of abilities will have to be extensionally adequate.

This adequacy condition suffices to identify some views of abilities as outright false. As Maier (2014) remarks, for instance, Aristotle discusses and rejects a view, according to which agents have the ability to do exactly what they actually do. The passage he cites is the following:

> There are some (...) who say that something is capable only when it is acting, and when it is not acting is not capable. For example, someone who is not building is not capable of building, but someone who is building is capable when he is building; and likewise too in other cases." (Makin 2006, 3, 1046b)

Aristotle rightly goes on to point out that "[i]t is not hard to see the absurd consequences of this" (ibid.). The grounds for his dismissal, I take it, is the adequacy condition I am suggesting; a view of abilities can only succeed if it is extensionally adequate.

What exactly does extensional adequacy amount to? We have to be careful not to misinterpret this condition. A very common misinterpretation would be to think that we need to have a clear and complete grasp of the extension of "ability" in order to be able to judge whether or not a view of ability is extensionally adequate. If this were so, one might reasonably doubt that the extensional adequacy of a view could ever be settled. For, arguably, we do not have such a clear and complete grasp of the concept of ability, or any other concept, for that matter. With respect to some of our ideas about abilities and some of our particular judgments about whether or not an agent in such and such a condition has an ability to φ, we may well be mistaken. Disagreement can be taken as an indicator here: whenever people disagree and the disagreement is not merely verbal, one of them has to be wrong. Apart from our misjudgments, there may be a grey area of cases to which our concept of ability may or may not apply. Having a concept need not necessarily imply that it be sharp enough to cover every single case. Philosophical theorizing may be partly normative: systematizing our thinking about a subject matter will often give us a reason for a particular sharpening of the concept within the grey areas.

What *does* extensional adequacy amount to, then? To three conditions: the clear cases of an agent having an ability have to be covered; the clear cases of an agent lacking an ability have to be excluded; and absurd consequences have to be avoided. This can be done without having a clear and complete grasp of the concept involved. All that is required is that we have a clear grasp on the applicability of the concept to *some* cases. I think we can confidently say that this condition is met in the case of abilities.

The first adequacy condition is not peculiar to a view about abilities. Extensional adequacy is a standard that any philosophical theory of any subject matter whatsoever will have to live up to. Let us now move on to adequacy conditions that emerge specifically in connection with abilities.

1.2 Adequacy condition 2: General and specific abilities

Throughout the literature on abilities, it is taken for granted that there are two different types of abilities. There are what we can call "general abilities" on the one hand, and "specific abilities" on the other.[1] Intuitively, the distinction is well-taken and the examples used in order to illustrate it all exhibit the same basic structure. Yet there is little systematicity in the way the distinction is being drawn; a lot of the pertinent ideas on the issue are implicit, rather than explicit, in people's writings, and many ideas surrounding the distinction are highly contested. In what follows, I will try and identify the core idea of the distinction. Later on in this book (→ 4.5), we'll turn to the more controversial issues and see how several ideas about general and specific abilities that float around in the literature relate.

The distinction between general and specific abilities is usually introduced by way of examples. Maier, for instance, asks us to

> [c]onsider a well-trained tennis player equipped with ball and racquet, standing at the service line. There is, as it were, nothing standing between him and a serve: every prerequisite for his serving has been met. Such an agent (...) has the specific ability to serve. In contrast, consider an otherwise similar tennis player who lacks a racquet and ball, and is miles away from a tennis court. There is clearly a good sense in which such an agent has the ability to hit a serve: he has been trained to do so, and has done so many times in the past. Yet such an agent lacks the specific ability to serve (...). Let us say that such an agent has the general ability to serve. (Maier 2014, 1.3)

Essentially along the same lines, Vihvelin points out that

> 'ability' is used in two different ways. Does someone with a broken leg have the ability to ride a bike? That depends. Speaking one way, we might agree that she has the ability. (She took lessons, she knows how; she has the necessary skills and competence.) Speaking another way, we might deny that she has the ability. (Her leg is broken; her bike-riding skills are temporarily impaired). (Vihvelin 2004: 448, fn.3).

Intuitively, the distinction emphasized in these two passages is clear enough. The point is simply that when we ask whether some agent has the ability to serve, for

---

[1] The distinction goes by a variety of names. Whittle (2010) speaks of local vs. global abilities, Berofsky (2002) speaks of token vs. type abilities. Nothing hinges on these terminological issues.

example, the answer may vary, depending on the sense of "ability" we have in mind. When we think about an agent's general abilities, we abstract away from the agent's present situation a great deal. It does not matter whether the agent lacks a racquet, or has a broken leg, etc. Thus, in the general sense of "ability",

> Walter may have the ability to walk even though he is bound to a chair. Sally may have the ability to sing even though she freezes whenever her Aunt is present. Chip may have the ability to cook even though no cooking equipment is available to him now. (Whittle 2010: 2)

More generally put, "[a]n individual with [a general] ability may be unable to exercise it in a particular case because a temporary obstacle is present" (Berofsky 2002: 196). As you can tell from these statements, there is a core understanding of general abilities, which is voiced over and over again by writers across the board: unfavorable temporary circumstances do not rob us of our general abilities.

Specific abilities are different. When we are interested in an agent's specific abilities, we do not abstract away from the agent's actual situation. If the agent has a broken leg or lacks a racquet, then this deprives her of the specific ability to ride a bike and play tennis, respectively. In the specific sense of "ability", one is unable to walk when tied up, to sing when nervous to the point of freezing, and to cook without any equipment at hand.

When people move away from particular examples and try to give a more general account of the distinction between general and specific abilities, they usually do so in much the same way in which I have framed the distinction in the last paragraph. Thus, Mele points out that

> a general practical ability (...) is the kind of ability to A that we attribute to agents even though we know they have no opportunity to A at the time of attribution and we have no specific occasion for their A-ing in mind. (...) [A] specific practical ability [is] an ability an agent has at a time to A then or on some specified later occasion. (Mele 2003: 447)

In very much the same spirit, Whittle distinguishes between "what an agent is able to do in a large range of circumstances, and what the agent is able to do now, in some particular circumstances" (Whittle 2010: 2). And again in similar terms, Vetter points out that general abilities are "retained through great changes in circumstances" (Vetter, ms.) and calls this feature the *robustness* of such abilities. What seems to be common ground, then, is that the having of a general ability is largely independent of particular features of the agent's situation, while the having of a specific ability is not.[2]

Note, by the way, that often no explicit distinction is made between a lack of opportunity and unfavorable internal features of the agent when philosophers think about factors that deprive an agent of a specific ability. The lack of a specific ability which Maier envisages involves the lack of a racquet and a ball and being absent from a tennis court. This looks very much like a lack of opportunity to play tennis. Vihvelin, in contrast, notes cases in which the agent's leg is broken or her bike-riding skills are temporarily impaired. These look more like unfavorable internal features of the agent in virtue of which the agent cannot ride her bike in her current situation.[3]

One may or may not like the fact that the notion of "specific abilities" that seems to prevail in the literature does not differentiate between external and internal conditions for establishing a lack of a specific ability. I will come back to this in chapter 4.5, b. For now, let us simply note that the having of a specific ability, as the term is commonly used, seems to require both: opportunity and the agent's being in a favorable condition herself.

From the fact that specific abilities do, but general abilities do not depend on particular features of individual situations, we can derive another insight about general and specific abilities, which is implicit in virtually everything that is written in connection with the distinction: general and specific abilities can, and often do, diverge. Thus, an agent may well have a general ability to ride her bike, say, and yet lack the specific ability to do so. This should be obvious from what has been said so far. The fact that someone has the ability to φ independently of any specific set of circumstances, or, as Whittle puts it, across a broad range of circumstances, does not imply that the same agent has the ability to φ in one very specific set of circumstances. Some obstacle may be present in those circumstances – the situation may be such that the agent cannot φ in that particular situation.

---

[2] It is the same core difference that Honoré has in mind, I believe, when he says that what he calls "can (particular)" is so labelled "because it is used primarily in connection with particular actions" while what he calls "can (general)" is so labelled "because it is most commonly used in connection with types of actions" (Honoré 1964: 463f.). Linguistically, Honoré's claim here may not be entirely sensible (cf. Palmer 1977: 3). The idea *underlying* it seems to be the very same thought one finds in all of the quotes we have looked at before, however.

[3] Note that Vihvelin (2013) uses the terms "narrow ability" and "wide ability" to distinguish between abilities that are had in virtue of intrinsic features (narrow abilities) and extrinsic features (wide abilities) of the agent.

What we know, then, is that there is a difference between general and specific abilities and that that difference has something to do with the fact that the having of specific abilities depends on the agent's particular situation, whereas the having of a general ability does not. A question that naturally comes up is: what does the having of a general ability depend upon, then? The answer to this question is usually that general abilities have something to do with what we can call the *modal reliability* of an ability: the agent has to be able to ɸ across a whole *variety* of situations. Whittle is very explicit on this point. For while she would agree that general abilities do not require that the agent be able to perform some act in a particular situation, she stresses that it does nevertheless require that the agent be able to perform *across* particular occasions. To put it in her terms, there has to be a large range of circumstances in which the agent is able to do what she is said to have a general ability for.

The same picture is outlined by a number of authors. Honoré, for instance, notes in connection with general abilities that

> [i]f a typist claims that she can type sixty words a minute, she is asserting (...) not merely that she has once managed by a fluke to get to sixty words in the minute. (Honoré 1964: 465)

Rather, what the typist commits herself to is something along the lines of the claim that "[s]he *normally* succeeds (...) when [s]he tries" (ibid., my emphasis). In the same spirit, Maier points out that

> [i]n order to have a general ability, it is not sufficient that one now have Aing as an option; rather it depends on one's relation to A in a wide range of circumstances. (Maier 2013: 11)

We can note at least three points, then. First, the having of a specific ability, but not the having of a general ability, depends crucially on the features of the agent's particular situation. Secondly, an agent can have a general ability without having the corresponding specific ability. Thirdly, the having of a general ability requires that the agent be able to ɸ across a range of different situations.

As you will note, then, we don't know much. Our short list of insights, while certainly useful in order to explicate our intuitive grasp of the distinction between general and specific abilities, is no more than a first step towards a firm understanding of the two notions. From a comprehensive view of abilities, we are entitled to expect a great deal more. How exactly do general and specific abilities relate, for instance? How do they fit into a view of abilities as a whole? How do the truth conditions of ability ascriptions vary, depending on the use of ability we have in mind? And can we nevertheless formulate uniform truth conditions for the having of an ability, if abilities come in the two distinct stripes we distinguished?

A comprehensive view of abilities, I take it, will have to provide answers to all of those questions. They emerge from the mere acknowledgement of the distinction between general and specific abilities, without presupposing anything specific about either of the two kinds of abilities or the relation between them. If there is a difference between general and specific abilities, then we will want to know what that difference is, and we will want to understand how the truth conditions for having the one and the other kind of ability relate. This gives us a second adequacy for a comprehensive view of abilities:

> A2. Any comprehensive view of abilities will have to elucidate the distinction between general abilities on the one hand and specific abilities on the other.

1.3 Adequacy condition 3: Degrees and context sensitivity

In the last section, we saw that abilities come in two kinds: there are general abilities on the one hand, and specific abilities on the other. Equally obvious, but very rarely acknowledged in the debate is a second feature of abilities: abilities come in degrees.[4] Glenn Gould has the ability to play the piano to a higher degree than any amateur pianist; my grandmother sews better, my cousin drives better, my father draws better than I do. A meteorologist who predicts the weather correctly in 9 out of 10 cases is better at predicting the weather than his colleague, who goes wrong every other time. The archer who hits the target every time she shoots is better than the archer who fails from time to time. We practice things to become better at them. We admire people who have outstanding abilities in some field or other. Degrees of abilities, I take it, are phenomena with which we are all most familiar.

Linguistically, we have various means to express relations between the degrees of two agent's abilities – ability ascriptions are gradable. We can say things like "Max can play the guitar better than I", or we can say even more explicitly "he has the ability to play the guitar to a higher degree". Typically, however, we use more subtle ways to grade ability ascriptions. We say "He plays the guitar better" or "He is better at playing the guitar", and in those statements the modal for ability is altogether dispensed with. Sometimes, we also use constructions like "He is a more able player". All of those kinds of statements are ways of

---

[4] This is a feature ability share with dispositions. For dispositions, gradability has been prominently noted by Manley & Wasserman (2008).

grading abilities – of making statements about the degree of an agent's ability in relation to one of her own or some other agent's or agents' abilities.

What this indicates is that, typically, the having of an ability is not an all-or-nothing matter, but rather a matter of having the ability to some degree or other. Perhaps there are exceptions. Perhaps there are abilities that do not seem to allow for gradation in any useful sense. Examples for such abilities are extremely hard to come by, though. Even abilities to perform very basic actions – the ability to push a certain button or to raise one's finger, for instance – seem to be gradable on closer examination. A child may have to develop enough strength to push a certain button, and the stronger she gets the better she gets at pushing it. A recovering stroke patient may get better and better at raising her finger. At least for general abilities, it is very hard (though perhaps not impossible) to think of examples that do not allow for degrees in any useful sense.

Does it make sense to think of specific abilities as coming in degrees as well? On the one hand, whether or not someone is able to perform a given action in a given situation seems to be an all-or-nothing matter. Either the agent can φ in a particular situation or she cannot. Maier, in one of the very rare passages on degrees of ability that can be found in the literature, subscribes to this very view when he writes that

> [i]n some given circumstances, an agent either does or does not have the option of Aing. In the case of general abilities, there are further distinctions to be drawn. For example, one agent may have the general ability to speak Mandarin, but another agent may have the general ability to a greater degree (she knows more phrases, is quicker to respond, and so forth). (Maier 2013: 12)

What Maier calls "options" are very closely related to what we have been calling "specific abilities" (→ 6.4). And options, Maier claims, are not gradable. Looking more closely, I doubt that the restriction of gradability to general abilities is correct, though.

Just think of two pianists whose general abilities to play a difficult piece are perfectly on a par. Now suppose one of them has had too much coffee this morning and is still a little shaky. In that case, it seems that that pianist has the specific ability to play the difficult piece here and now to a lower degree than the other one. The pianist who does not suffer from a coffee overdose is better at playing the piece than the overdosed pianist. It is not that the overdosed pianist lacks the specific ability altogether. She *can* play the piece here and now. The problem is rather that her present condition does not allow for as good of a performance as the one the other player is able to deliver in the same situation. It thus seems that she has the specific ability, albeit to a lower degree.[5]

Which factors play a role in determining an agent's degree of an ability? This question, though quite natural in the face of the obvious gradability of abilities, is not touched upon in the literature very often. However, Millikan offers helpful guidance here when she distinguishes between two senses of "learning to do something better". In a first sense, learning something better can "consist in learning how to turn out a more polished product" (Millikan 2000: 55), as in the statement "She learns to play the violin better". In a second sense, "you can learn to do the same thing under more circumstances. For example, you may learn to drive a car safely even on ice" (ibid.).[6]

This suggests that abilities can be graded along at least two dimensions. On the one hand, there is what we can call "the dimension of achievement". Here, what matters is the quality of the performance the agent is able to deliver. The higher the quality of the performance the agent is able to deliver, the higher is, *prima facie*, the degree of the agent's ability to φ.

Examples are easy to come by. The singer who sings perfectly on every occasion is clearly a better singer than the agent who sings poorly every time. The weather forecaster who gets the weather conditions exactly right each time is better than the one who gets them only roughly right. A barista who serves you a perfectly crafted coffee is better at making coffee than an average waiter.

On the other hand, there is what we can call "the dimension of reliability". Here, what matters is the range of circumstances across which the agent manages to deliver a performance. The archer who hits the target every single time she shoots is obviously better at hitting the target than the archer who misses every other time. And the weather forecaster who is right in most cases is better at predicting the weather than her colleague who goes wrong most of the time.

Note that it is not important that the circumstances across which an agent achieves higher reliability differ substantially from one another. Millikan's example of someone learning to drive on ice, who was unable to do that before, may

---

[5] Maier responds to these considerations by acknowledging that there may be "some other notion which is similar to options but for being gradable". He goes on to argue, though, that "we already have a non-gradable notion, and that is the one which I am picking out with the term "option"" (ibid.: n13).
[6] Actually, Millikan proposes a third sense, which is also strongly emphasized by Sosa (2015): the sense in which you get better at something by "getting better at recognizing risky situations, getting better at knowing when not to try, or not to try this way but rather to try some other way" (Millikan 2000: 56). In order to keep things reasonably simple, I will neglect this third sense in what follows.

suggest this. But in fact, an agent also becomes more reliable in doing something the more reliably she succeeds in doing that thing in the very same kinds of circumstances.

The dimension of reliability and the dimension of achievement are usually not considered in isolation. When we grade an agent's overall ability to perform some given act, we set the dimensions off against each other. Which dimension carries which weight is a matter of the particular ability under consideration and also of the context in which the grading takes place.

An archer's ability to hit a target is usually only evaluated on the dimension of reliability. The more reliably the archer hits the target, the better she is at hitting it. The piano player's ability to play a certain piece is usually only evaluated on the level of achievement. The player who is able to deliver the better performance is usually counted as the better player even if she can perform in far fewer circumstances than another player.

In most cases, we take both dimensions into account and set them off against each other in ways determined by context. The weather forecaster's ability to predict the weather, for instance, is exhibited to a higher degree the more precise her prediction of the weather is across the more situations. In cases of that kind, an increased level of reliability can outweigh a comparably low level of achievement and *vice versa*.

The inherently gradable nature of many abilities is a particularly important aspect within a theory of abilities, because it seems to underlie yet another crucial feature that will have to be captured by such a theory: ability ascriptions exhibit a kind of context sensitivity that is generally exhibited by statements involving gradable expressions. Take a gradable adjective like "tall". "Tall" is gradable – Elsa is tall, for instance, but Max is a little taller. The same goes for "rich", "flat", "dark", "sweet", "boring", and a long and well-known list of further adjectives. The gradability of such expressions reflects an important feature of the properties they pick out: richness, flatness, darkness, sweetness, boringness, and the like all come in degrees. Grading gradable adjectives is a way of saying that the property they express is exhibited to a higher degree in one case than in another.

Importantly, gradable adjectives are context sensitive (Kennedy 2007). And even more importantly, their context sensitivity emerges directly from their gradability and the underlying fact that the properties they express come in degrees. The reason for that is easy to grasp, and we may focus on "tall" as an example. While instances of tallness, say, are simply heights arranged along a scale in ascending order, the gradable adjective "tall" ascribes tallness *simpliciter*. And which property is that? That varies across speaker contexts. If we talk about a particularly tall two-year-old, "tall" will pick out a very different property than when we talk about a particularly tall basketball player or, even more to the point, a particularly tall building.

What we have is a scale of heights, or different degrees of tallness, if you will, and what the context does is fix a threshold on that scale above which the predicate applies *simpliciter*. That is just to say that "tall" and statements involving that predicate are context sensitive: the truth conditions for "x is tall" vary across contexts. In fact, the context sensitivity that is exhibited by gradable predicates is of a very specific kind: it is a sensitivity to what is often referred to as *standards* – they are sensitive to what speakers consider a sufficient degree on a scale.

Context sensitivity is most obvious when one and the same statement is true in the mouth of one speaker, but false in the mouth of another speaker. When I think of Lasse's height in comparison with the heights of his kindergarten friends, I may truly say "Lasse is tall". But when you think of the way he desperately tries to reach the window sill, you may truly say "Lasse is not tall". Those two utterances seem contradictory on the surface, but in fact they are not: I pick out different properties by "tall" across the two statements and therefore express perfectly compatible propositions by them.

Coming back to ability ascriptions, we can note that quite the same phenomenon can be observed. Someone who has driven a car once or twice and manages to get from A to B without running someone over, but drives extremely poorly otherwise, may be counted as having the ability to drive in one context, but not in another, for instance. When asked whether she can drive, she may truly say "yes" when a car needs to be moved from one space in an empty parking lot to another, say, but she may truly say "no" when a car needs to be moved from one neighborhood of a crowded city to another.

What this illustrates is that in different contexts, different degrees of the ability to drive are required in order for us to be willing to count someone as having the ability *simpliciter*. In the context in which a car needs to be moved on the empty parking lot, a very low degree of ability is required – pretty much any kind of driving is good enough for the purpose at hand in that context. In the second context, in which a car needs to be moved across town, in contrast, not any kind of driving will do. To count as having the ability to drive in that context, the degree of the ability to drive has to be high enough to make it through the city without running someone over, for example.

Ability ascriptions seem to exhibit a structure akin to ascriptions of other gradable predicates, then. The degrees to which the agent has the property picked out by the predicate form a scale, and context sets a threshold on that scale in order to determine which degree has to be exhibited in order for the predicate to apply *simpliciter* in the context.[7] For this reason, a lot hinges on a viable account of degrees. Once we have degrees, the corresponding kind of context sensitivity comes for free. Without degrees, the context mechanism for ability ascriptions remains mysterious as well.

This yields a third adequacy condition for a comprehensive view of abilities.

> A3. Any comprehensive view of abilities will have to account for degrees of abilities and the corresponding kind of context sensitivity that attaches to ability ascriptions.

1.4 Adequacy condition 4: Agentive vs. non-agentive abilities

Abilities are paradigmatically abilities to perform actions. When we think of examples of abilities, what immediately comes to mind are abilities to do things like playing the flute, picking bike locks, reciting a poem by heart, and the like. Let's call an ability to perform an action an "agentive ability". Agentive abilities are clearly the most paradigmatic cases of abilities.

On closer inspection, however, the relation between abilities and actions is not quite as tight as it seems at first sight. Surely, we want to say that most people have the ability to smell, see, and taste, for instance. But having a sense perception is obviously not an action. In smelling, seeing, and tasting the agent remains pretty much passive. Having sense perceptions is a suffering, to employ an old-fashioned term, rather than an action. Hence, it seems as though there are things agents can have an ability for which are not intentional actions. For a lack of a better term, let us call those entities "mere behaviors" and let us call an ability to show those mere behaviors "non-agentive abilities".[8]

Non-agentive abilities include abilities of various kinds. We already talked about abilities to have sense perceptions. But non-agentive abilities also include abilities to undergo a range of biological processes. Just think of abilities like digesting, producing saliva, adapting the size of one's pupils to the lighting conditions, sweating, sensomotorically tracking one's body parts, and so on. Most humans have those abilities; some don't. "She has lost her ability to digest food", "He is no longer able to produce saliva", and "She cannot track her body parts anymore" are perfectly reasonable denials of abilities. But again, the performances being talked about are not actions.

---

7 In fact, things are slightly more complicated. The context does not only set a threshold on the scale, but also determines the ordering on that scale (cf. Heller 1999).
8 Note that some authors may want to reserve the term "ability" for what I am referring to as "agentive abilities" and refer to non-agentive abilities in some different way (Maier 2014, 1.2).

I have heard people say that there is a derived sense of "intentional action" in which the having of a sense perception and the undergoing of biological processes can be performed as actions. All I have to do is bring the stimulus about that gets the process started. To digest, eat. To hear, expose yourself to sound. In such cases, the digesting and the hearing may be counted as intentional actions. Are they really, though? To me, counting those behaviors as actions sounds quite bizarre. But I guess a lot will depend on one's systematic goals in using the notion of an action here. Clearly, digesting and smelling are not actions in the paradigmatic sense of the word. Once started, the agent does not have any control over the process, she cannot abort it, she does not govern it. But perhaps some people will still want to count some instances of digesting and smelling as actions.

Luckily, nothing much hinges on this question. What is important for our purposes here is a different thing: non-agentive abilities *can be* and typically *are* exercised without a foregoing intention to exercise them. They *can* thus be exercised by being engaged in a mere behavior, which is clearly not an action. The agent who digests without intending to do so exercises her ability to digest, but her digesting is clearly not an action. The agent who hears without a foregoing intention exercises her ability to hear, but her hearing is clearly not an action, either. Even *if* non-agentive abilities may sometimes be exercised intentionally (which I am very hesitant to accept), all that matters for my purposes here is that they need not be exercised intentionally, and typically aren't.

Having said this, let's note that there is a more extreme class of non-agentive abilities as well. There are also abilities which *cannot* be exercised intentionally at all – not even in the very peculiar sense in which one may be able to hear or digest intentionally. The most obvious examples for abilities of that kind are abilities which can for conceptual reasons only be exercised without a foregoing intention. Examples are the ability to read street signs or advertisements *unintentionally* (Löwenstein 2013, 2016), to understand a word in a foreign language (van Inwagen 1983: 8–13) or follow a conversation at a different table *without wanting to*, or to recognize a familiar face in a crowd or a pattern of numbers *without actively striving for recognizing it*. These abilities wear their status as non-agentive abilities on their sleeves, and in contrast to the instances of non-agentive abilities we looked at in the beginning, they can, in fact, *only* be exercised if no intention to exercise them precedes them.

To see this point entirely clearly, note that the ability to read street signs is not the same ability as the ability to unintentionally read street signs. A first grader may already have the former ability – she can read street signs alright, but she has to put a lot of effort into it. Most grown-ups, however, also have the latter ability – reading street signs is something that just happens; they have acquired the ability to read them unintentionally.

What we can note, then, is that there are two very different classes of abilities. On the one hand, there are the paradigm cases of abilities: abilities to perform intentional actions. On the other hand, there are abilities like the ability to digest, read street signs unintentionally, or dissociate. In other words, there are abilities which need not, or even cannot, be exercised as actions. Abilities of that kind are what I have called "non-agentive abilities". From the distinction between agentive and non-agentive abilities we can derive a fourth adequacy condition for a comprehensive account of abilities:

> A4. Any comprehensive account of abilities will have to account for agentive as well as non-agentive abilities.

1.5 Explanatory challenge 1: Abilities and counterfactuals

So far, I have stated a set of adequacy conditions for a comprehensive view of abilities – conditions that will have to be met by any view that aims at capturing its subject matter appropriately. The adequacy conditions are of crucial methodological importance for what follows. They will serve as guide rails for the critique of existing accounts and for the construction of a superior view.

Let us now proceed to some explanatory challenges that arise once we have a comprehensive view of abilities on the table. Let us formulate, that is, a bunch of questions that arise quite naturally with respect to abilities and for which we can hope to get an answer from a comprehensive view, once we have it.

The first explanatory challenge arises from the insight that abilities seem to be somewhat closely related to certain counterfactuals. Typically, if I have the ability to φ, it is also true that I would φ if I tried, chose, desired, intended ... to φ. The exact formulation of the counterfactual need not concern us right now. The important point is that if an agent has an ability to perform some given action, then her performance of the action will typically be counterfactually linked to some suitable motivational state of hers, and *vice versa*. Normally, it holds that if I can whistle then I would whistle if I intended to. *Vice versa*, it normally also holds that if I would whistle if I intended to whistle then I have the ability to whistle.

In the debate on abilities, there is a strong emphasis on counterfactuals. In fact, many have taken the link between abilities and counterfactuals of this sort to be of the strongest possible kind: according to what we will later get to know as the *simple conditional analysis* (→ 2), for an agent to have an ability is precisely for some such counterfactual to be true. In the course of chapter 2, it will become clear that the simple conditional analysis exhibits some severe shortcomings. The link between counterfactuals and abilities is way more intricate than the simple conditional analysis would have us believe. Yet, it is hard to dispute that there is an interesting link between the having of an ability and counterfactuals of some sorts. And in fact, this link emerges directly from two essential features of abilities themselves.

The first of these two features is that abilities are modal properties in the sense that there is a difference between their instantiation and their exercise, and they need not actually be exercised in order to be actually instantiated. For some abilities this is very obvious: I have the ability to run around a tree with a sombrero on my head while singing the national anthem, for instance, but so far I have never exercised that ability and presumably I never will.

The second crucial features is that for every ability – even for strikingly weird ones like the one I have just mentioned – there have to be certain conditions which would bring their exercise about. If I have the ability to do a handstand, say, there has to be some possible world in which I do a handstand. For what would it be for me to have the ability if there is absolutely no possibility in which I could actually bring the action about? The performance of the act would be metaphysically impossible. And since it seems clear that no one has the ability to do what is metaphysically impossible, there has to be a world in which I actually do a handstand.

Given the standard semantics for counterfactuals (Lewis 1973; Stalnaker 1968), this feature alone suffices to show that for every ability there has to be *some* true counterfactual with the performance of the act in the consequent. If I have the ability to run then there has to be a true counterfactual of the form "If C were the case, then I would run". All that has to be the case is that C specifies conditions which are such that the closest worlds in which *those* conditions obtain are also worlds in which I run. In the worst case, C will not be a very interesting set of conditions – C may simply pick out a world in which the subject runs by specifying it completely. If we are lucky, however, the counterfactual will be more interesting in kind and show a counterfactual dependence relation, not just between entire world states and the agent's performances, but between some motivational state of the agent and her performances, say.

So there are two factors that constitute the link between abilities and counterfactuals; their modal nature on the one hand, and the fact that certain conditions would bring the exercise about on the other. What kind of counterfactuals will turn out to be interestingly related to abilities and whether counterfactuals play any important role in the formulation of an analysis of abilities remains to be seen (→ 6.2). It is definitely something we can wish to learn more about from a comprehensive view of abilities, though. Let's formulate this as our first explanatory challenge for a comprehensive view of abilities, then:

> E1: From a comprehensive view of abilities, we may hope to learn how abilities relate to counterfactuals.

1.6 Explanatory challenge 2: Abilities and dispositions

The next feature of abilities I want to draw attention to has received a lot of recognition recently: abilities resemble dispositions in many interesting ways.

Dispositions are properties like the fragility of a vase, the solubility of a lump of sugar, or the irascibility of a person. In recent years, people have become more and more interested in dispositions. That is because, just like abilities, dispositions are modal properties – there is a difference between their instantiation and their manifestation, and obviously, they can be instantiated without being manifested. A vase can be fragile without actually breaking. A lump of sugar can be soluble without actually dissolving. A person can be irascible without actually losing her temper. For this reason, it has seemed promising to many to relate dispositions to all kinds of other modally laden phenomena such as lawhood (e.g. Bird 2005, 2007), essentiality (e.g. Harré & Madden 1975; Mumford 2004; Bird 2005, 2007; Whittle 2008), and causality (e.g. Mumford & Anjum 2011) on the one hand, and to core modal notions such as necessity and possibility (e.g. Vetter 2015) on the other. The views that have emerged in the course of this debate are as manifold as in any other philosophical debate. But there seems to be a broad consensus that an understanding of modal phenomena in general will also require an understanding of dispositional properties.

The increased interest in dispositions has also shaped the contemporary debate on abilities. There has always been some tendency to assume a very close tie between the two phenomena (e.g. Ryle 1949; Moore 1912; Vihvelin 2004, 2013; Fara 2008; Smith 2003). The more elaborate the understanding of dispositions has become, the more similarities between dispositions and abilities have become visible. And the more of those similarities have become visible, the more explicitly have some authors tried to treat abilities as some particular kind of dispositions and to account for the former by whatever means will turn out appropriately to account for the latter.[9]

Apart from their shared status as modal properties, there are at least two further important features that dispositions share with abilities. First, both disposi-

---

[9] This doctrine is particularly stressed by so-called new dispositionalists (→ 5.4, 6).

tions and abilities are quite intimately related to counterfactuals; and not just that: the counterfactuals to which they are related structurally resemble the counterfactuals that seem to play an important role in connection with abilities. For fragile things, for instance, it will often hold that they would break if they were struck. And for soluble things, it will often hold that they would dissolve if they were placed in a liquid. Compare that with counterfactuals that seem to hold for agents having abilities: for an agent who is able to do a handstand, it will often hold that the agent would do a handstand, if she intended to. And for an agent with the ability to whistle, it will often hold that the agent would whistle, if she intended to whistle. The structural resemblance of all of those counterfactuals is obvious: for dispositions as well as abilities, the conditional posits a counterfactual dependence relation between what we can call a typical *stimulus condition* and the manifestation of the modal property. In the case of abilities, the stimulus that comes to mind is the agent's intention (or some other motivational state). In the case of dispositions, the stimuli will typically be external events.

In the last section, I said that the simple conditional analysis of abilities exhibits some severe shortcomings (→ 2). Interestingly, however, this does not result in an asymmetry between abilities and dispositions. Rather, it strengthens the similarities between the two. For as we will see later on (→ 6.2) the link between dispositions and the counterfactuals that link *their* stimuli to the relevant manifestations seems to be just as complicated, and partly for the same reasons. It is even more tempting, therefore, to think that abilities may just *be* dispositions of a certain sort.

The second important similarity between dispositions and abilities has to do with the role of degrees and context sensitivity. Just like abilities, dispositions come in degrees (Manley & Wasserman 2008). A champagne glass is more fragile than a coffee mug or a brick, Klaus Kinski was more irascible than most Buddhist monks, an aspirin is more soluble than honey. Moreover, disposition ascriptions seem to be context sensitive: "engineers working with slabs of concrete and chemists working with glass electrodes may ascribe different properties with the predicate 'fragile'" (ibid.: 76).

As in the case of abilities, the two points seem to be tightly related; again, it is plausible to think of the context sensitivity of disposition ascriptions in terms of degrees: to take fragility again as an example, it seems to be the case that varyingly high degrees of fragility will be required in different contexts for the term "fragile" to apply *simpliciter*. In the chemists' context, for instance, a higher degree of fragility will be required than in the engineers' context.

So both dispositions and abilities are modal properties, seem to be somehow connected to counterfactuals of roughly the same form, both come in degrees, and ascriptions of both therefore exhibit the same kind of context sensitivity. In view of those similarities, the thought that abilities simply are dispositions suggests itself very naturally.

Yet, the exact relationship between abilities and dispositions is far from obvious on closer inspection. Surely, the ability to do a handstand is not simply the disposition to do a handstand; obviously, I can have the ability without having the corresponding disposition. The same goes for "the ability to kill cats, to stand on my head on the commuter train, and to play bebop on the violin" (Millikan 2000: 52). If abilities are dispositions at all, it therefore has to be a different, and presumably a more subtle one.

Moreover, there seems to be an important difference between abilities and dispositions which *prima facie* speaks against the idea that abilities *are* dispositions: abilities have an obvious active component, whereas dispositions seem to be passive in kind. When we think of paradigmatic cases of dispositions – solubility, breakability, irascibility – we will note very quickly that the manifestation of that disposition is something which the subject undergoes, rather than exercises. In the case of abilities, this is very different. Here, the paradigmatic cases – think of the ability to sing a song, to build a shag, to play tennis – all have an action as their manifestation: the agent controls what is going on and she also controls whether to exercise the ability at all. All this raises an explanatory challenge for an account of abilities:

> E2. From a comprehensive view of abilities, we may hope to learn about the relation between abilities and dispositions.

1.7 Explanatory challenge 3: Abilities and "can" statements

When we want to ascribe an ability to an agent, we have quite a few linguistic means available. On the one hand, we can say "S has the ability to ϕ" or "S is able to ϕ". On the other hand, though, we can say "S can ϕ" and ascribe the ability to ϕ via what I will call a *"can" statement* – a statement involving the modal auxiliary "can". Thus, we can say things like "Jane can cook", "John can do a summersault", and "Jack cannot run very well". What we express via those statements is that Jane, John, and Jack each do or do not have a certain ability – the ability to cook, to do a summersault, and to run, respectively.

Importantly, that is not to say that all "can" statements are ability ascriptions, though. "Careful, you can trip here", "You can't do that!", and "Ann cannot be the murderer" are but a few examples of "can" statements which clearly do not express an ascription or denial of an ability. What we can note, then, is

1.7 Explanatory challenge 3: Abilities and "can" statements

that abilities can be ascribed via "can" statements of a particular kind. There is what we may call the *"can" of ability*.

Note that the use of "can" statements is not restricted to the ascription of either general or specific abilities. A brief look into the Corpus of Contemporary American English (COCA)[10] gives us a variety of real-life examples in which "S can ϕ" (or its negative "S cannot ϕ") is used to express that S has (or lacks) some ability to ϕ. And as is easy to see, there are examples for both "can" statements which express general, and "can" statements which express specific ability. Here are a few examples in which a "can" statement is used to ascribe (or deny) a *general* ability – what is said is that some agent has, independent of their actual situation, the ability to perform some action.

1. At least I am young and strong, and I **can swim**.
2. You **can walk** and chew gum at the same time.
3. Oh, yeah. Some people **can not dance**.
4. She **can not cook**, she says, but she's learning.

Examples for the use of a "can" statement by means of which a specific ability is ascribed are just as easily to come by:

1. You're wrong, your honour. You **can stop** this right now before you do it.
2. There is nothing you **can do** about it now.
3. "He's settling down. You **can handle** him safely now."
4. People are measured only by what they **can do** here and now, immediately.

In examples 10 – 15, "can" ascribes a specific ability – the can-statement expresses that the agent has the ability to perform some given action in some given situation. This is indicated very clearly by the use of expressions such as "now" or "here and now", which serve the purpose of picking out a specific situation. We can note, then, that abilities in general can be ascribed via "can" statements.

That abilities can be ascribed via "can" statements is a very interesting and far-reaching insight. According to the standard view on "can" statements (Kratzer 1977, 1981), "can" statements obey a uniform semantics. And according to that semantics, all "can" statements express restricted possibility. On that view, then, to say p can be the case is only to say that it is possible, in a properly restricted sense, that p be the case.

That ability ascriptions can be expressed via "can" statements is thus a far-reaching insight, because either the standard semantics is true and ability ascriptions, too, express restricted possibility, or ability statements do not express

---

[10] http://corpus.byu.edu/coca/

restricted possibility and the standard semantics does not meet its claim to universality – it would be false when it comes to the "can" of ability.

Thus, we get an interesting result, no matter which of the two possibilities turns out right. If ability statements do indeed express restricted possibility, then this takes us a very long way towards a comprehensive view of abilities. The basic structure of the view would be settled: S can ϕ, on that view, if and only if it is possible, in the properly restricted sense, for S to ϕ. What would remain to be seen is what kind of restrictions are the relevant ones.

If ability statements do *not* express restricted possibilities, however, then there is something wrong with the standard semantics. The "can" of ability would behave differently from what the standard semantics predicts. In that case, it would have to be shown in detail how the semantics of ability statements differs from the standard semantics for "can" statements.

Whatever one thinks about abilities, then, one had better have some view on the connections between ability ascriptions and the standard semantics for can-statements. This can be formulated in terms of an explanatory challenge for an account of abilities:

> E3. From a comprehensive view of abilities, we may hope to learn how ability ascriptions relate to the semantics for "can" ascriptions in general.

1.8 Upshot

In this chapter, we have identified four adequacy conditions for a comprehensive view of abilities and three explanatory challenges. The adequacy conditions will serve as guide rails for the discussion to follow: they provide conditions that will have to be met by any view of abilities that sets out to capture its subject matter appropriately. Here they are in a nutshell.

Any comprehensive view of abilities will have to...

- A1. ... be extensionally adequate.
- A2. ... elucidate the distinction between general and specific abilities.
- A3. ... account for degrees of abilities and the corresponding kind of context sensitivity that attaches to ability ascriptions.
- A4. ... account for agentive as well as non-agentive abilities.

The explanatory challenges, in contrast, point to questions that apply only after having settled on a comprehensive view of abilities. They point to questions towards which one can expect a comprehensive view of abilities to take a stance. Here they are.

From a comprehensive view of abilities, we may hope to learn...

E1: ... how abilities relate to counterfactuals.
E2. ... about the relation between abilities and dispositions.
E3. ... how ability ascriptions relate to the semantics for "can" ascriptions.

We are now in a position to dive into the discussion of the two most prevalent views in the literature. The next chapter turns to the so-called "simple conditional analysis of abilities". Chapter 3 turns to what I will refer to as "possibilism".

2 The simple conditional analysis

This and the next chapter are all about learning from others. And by that I mean not just learning from their mistakes, but also learning from their insights. What I will do, therefore, is take a detailed look at the two most prevalent traditional views of abilities, extract their most appealing insights, and argue why none of them is suited to provide a comprehensive view of abilities.

The first view I will discuss, and the one that will concern us in this chapter, is what I will refer to as the "simple conditional analysis of abilities". On that view, having an ability is a matter of a certain counterfactual being true. In its most prominent version, the view states that an agent has an ability to ϕ if and only if the agent would ϕ, if she so chose (Moore 1912, ch. 7). The view can be traced back to Hume, who famously held that

> [b]y liberty, then, we can only mean a power of acting or not acting according to the determinations of the will; that is, if we choose to remain at rest, we may; if we choose to move, we also may. (Hume 1748, 8.23/ 95)

The simple conditional analysis used to be very popular among compatibilists in the free will debate, who, like Hume and Moore, saw it as a means to provide a compatibilist position on free will and determinism. But it was also met with sympathy by many logical positivists, including Schlick (1939) and Ayer (1946).

Different versions of the view differ with respect to the motivational state figuring in the antecedent of the counterfactual. What all of the views have in common, however, is that they analyze abilities in terms of a single counterfactual conditional, according to which the agent would have performed an action, if she had been properly motivated to perform it.

I am emphasizing that the simple conditional analysis works with just one counterfactual, because we will see in chapter 6.2 that there are more sophisticated ways of analyzing abilities by employing counterfactual conditionals. Instead of using just one counterfactual, the views we will be looking at in that section work with a large quantity of such conditionals, each of which has a highly specific antecedent. They can therefore be seen as a refined successor of the simple conditional analysis. I will subsume them under the label "the sophisticated conditional analysis" later on.

Since the sophisticated conditional analysis avoids very many of the problems that are going to be brought up for the simple conditional analysis in this chapter, I want to emphasize very clearly that this chapter deals exclusively with the simple conditional analysis and its problems. It deals with views that analyze abilities in terms of a single counterfactual conditional of the form "If

S had been properly motivated to φ, S would have φ'ed". Seeing the limitations of that kind of view will be extremely helpful in seeing how to do better. And at this point in the book, this is just the kind of insight we are after.

For simplicity and vividness, I will concentrate on one particular version of the simple conditional analysis in what follows. And for reasons of uniformity with my own account of abilities, that version will be one which posits a counterfactual connection between the agent's *intentions* and her performances. More precisely, I will focus on the view that

SCAA. An agent S has an ability to φ if and only if S would φ, if S intended to φ.

I would like to emphasize, however, that none of the arguments that are going to be developed in this chapter hinge on the differences between this version of the simple conditional analysis and any other version of the view. Everything I say in this chapter applies to any analysis of abilities in terms of a single counterfactual, which ties some kind of motivational state of the agent to her corresponding performances.

I think we can say without reserve that SCAA is an extremely appealing view at the outset. Abilities, we know, are modal properties – there is a difference between their instantiation and their manifestation, and the former does not require the latter. For an agent to have the ability to do a handstand, say, the agent need not actually do a handstand. Yet, of course, the ability to do a handstand clearly has *something* to do with the agent's effectively doing handstands; if not in the actual circumstances, then surely in possible circumstances. The conditional analysis captures this thought rather elegantly: to have the ability to do a handstand, the agent need not actually do a handstand; what she needs to do is do a handstand when intending to do one. That seems like a very reasonable requirement.

In view of the initial plausibility of the view, it is unsurprising that SCAA is still the default view in many areas of philosophy and the view many try to cling to when encountering objections. The idea that abilities will turn out to be a matter of *some* counterfactual or other being true is still very deeply entrenched in the contemporary debate.

In what follows, I will argue that this idea is mistaken. The claim I will establish in the course of this chapter is that abilities are not properly accounted for in terms of SCAA, no matter how exactly the counterfactual is spelled out. SCAA faces a number of structural problems, which together amount to the failure of the view, and it does so regardless of the details of any specific version of the account.[1]

The structure of the section is as follows. Since we are dealing with counterfactuals a lot in this chapter, I will start, in the next section (2.1), by elaborating on the semantics of counterfactuals generally. This will provide us with a crisp terminology and a firm grip on the lines of argument to come. Readers familiar with the standard semantics for counterfactuals are well-advised to skip that section.

The subsequent sections are devoted to the critique of the simple conditional analysis. As I will argue, the analysis fails because it runs into problems with all of the adequacy conditions for a comprehensive view of abilities. The view is extensionally inadequate, because it runs into problems with masks (→ 2.2) and cases in which the agent cannot form the intention to ɸ to begin with (→ 2.3), it fails to provide an account of general and specific abilities (→ 2.4), it fails to offer an account of degrees of abilities and the corresponding kind of context sensitivity that attaches to ability statements (→ 2.5), and it fails to provide an understanding of non-agentive abilities (→ 2.6). By the end of this chapter, it should be clear that the simple conditional analysis, while highly appealing at the outset, is deeply flawed. Abilities are not a matter of the truth of a counterfactual.

2.1 Modal semantics: Counterfactuals

A *counterfactual*, as I use the term, is a modal statement in the form of a subjunctive conditional – it states that p would be the case, if q were the case. This is a modal statement insofar as p, and thus q, need not be the case for such a statement to turn out true. Most counterfactuals have an antecedent which specifies a non-actual event; thus, the name "counterfactual". Despite the name, however, this is not essentially so. Some subjunctive conditionals specify an actual even in their antecedent, and they typically also go by the name "counterfactual".[2] I'll stick to that practice throughout the book.

Let's get some counterfactuals before us. Here are some true ones. If I had gotten up at 6, I would have heard the church bells; if you were the president

---

[1] Note that the criticism I am going to develop is not meant to provide an exhaustive assessment of problems that beset the simple conditional analysis. There is, for instance, a huge amount of literature on the "iffyness" of ability statements, which took its starting point with Austin's (1956) famous criticism of the simple conditional analysis, which I will not go into here.
[2] This is consistent with Lewis's (1973) use of the term.

of the United States, you would be more powerful than you are now; if birds didn't have feathers, they would be cold in the winter. Here are some false ones. If I had gotten up at 8, I would have heard the church bells; if you were the president of the United States, you would be less powerful than you are now; if birds didn't have feathers, they would wear little knit sweaters.

What makes counterfactuals true and false, respectively? The standard semantics (Lewis 1973; Stalnaker 1968) states the truth conditions of a counterfactual in terms of possible worlds. Possible worlds are complete ways things can be – "single, maximally inclusive, all-encompassing situation[s]" (Menzel 2016).

Theoretically, possible worlds can be thought of as a variety of things. Lewis famously thought of them as real entities, viz. entities of the same kind as the actual world, but causally, temporally, and spatially completely unrelated to it (Lewis 1986). Most philosophers, however, do not think of them as material entities. Usually, possible worlds are thought of as abstract objects (Plantinga 1974, 1976), as sets of propositions (Adams 1974; Fine 1977) or as "useful theoretical entities having no independent reality" (Menzies 2014) at all. I am not going to commit on this issue throughout this book; you are free to make your pick.

We can now introduce a very important notion: the notion of *similarity*, or *closeness*. Possible worlds resemble each other with respect to what is true in them. They resemble each other in certain aspects, but also overall. And this allows us to impose an ordering on them. This ordering can be thought of in spatial terms:

> [I]magine all possible worlds spewed out across the logical universe, with the actual world in the center [and suppose that] the relative distance from a given world to the actual world is a measure of the relative similarity of that world to the actual world. (Heller 1999: 116).

The more similar a world is to the actual world, the closer it is to the actual world in modal space. The more dissimilar, the further away. This is all the apparatus we need to state the truth conditions for counterfactuals. Because we can now state, with Lewis (1973: 16), that

> STANDARD-1. "If p were the case, q would be the case" is true at a world w if and only if (i) some p-world where q holds is closer to w than any p-world where q does not hold or (ii) there is no possible p world, in which case the counterfactual is vacuously true.

Counterfactuals with impossible antecedents need not concern us in this chapter, so let's ignore condition (ii). The important condition for our purposes here is condition (i): the condition that some p-world where q holds is closer than any p-world where q does not hold. Complicated as the condition seems at first sight, the idea underlying it is actually very simple and intuitive. It is just that "a counterfactual is true, if it takes less of a departure from actuality to make the antecedent true along with the consequent than to make the antecedent true without the consequent" (Menzies 2014).

Apply this to an example. Why wouldn't birds wear sweaters, if they didn't have feathers? The answer provided by STANDARD-1 is: because it is not the case that some world in which they lack feathers, but wear sweaters is closer than any world in which they lack feathers and don't wear sweaters. Many worlds in which they lack feathers and don't wear sweaters are much closer.

Despite the fact that (i) is quite intuitive on second glance, it is obviously a bit unwieldy. I will therefore help myself to a modification of the condition, which we gain if we accept what is known as the *limit assumption*: the assumption that there are always closest p-worlds.[3] Plausible at first, the limit assumption is actually very controversial, and Lewis takes it to be false. Take a counterfactual with the antecedent "If I were taller". The problem with such a counterfactual is that it is hard to see what the closest worlds are in which I am taller. Since height is continuous, there will be an infinity of increasingly similar worlds between the most similar world in which I am exactly one centimeter taller and the actual world. Which is the most similar world in which I am taller (at all), then? It seems like there is none.

To deal with the possibility of ever more similar worlds, Lewis formulated the truth conditions of counterfactuals along the lines of STANDARD-1. Neglecting problems with the limit assumption, however, one can simplify the first condition quite a bit and restate the truth conditionals for counterfactuals as follows:

> STANDARD-2. "If p were the case, q would be the case" is true at a world w if and only if (i) the p-worlds that are closest to w are q-worlds or (ii) there is no possible p world, in which case the counterfactual is vacuously true.[4]

Instead of postulating that some world in which the antecedent *and* the consequent holds be closer to w than any world in which the antecedent is true, but the consequent false, our new condition (i) simply postulates that the closest worlds to w in which the antecedent is true are worlds in which the consequent is true as well. The new condition implies the earlier condition, given that there

---

[3] One of the central differences between the otherwise very similar views of Lewis and Stalnaker on counterfactuals is that Stalnaker accepts, while Lewis denies the limit assumption (Lewis 1973: 78).

[4] This comes very close to Stalnaker's version of the semantics.

are closest antecedent-worlds, which is just what the limit assumption presupposes.

In what follows, I will act as though the limit assumption were unproblematic and use STANDARD-2 whenever I state or refer to the truth conditions for counterfactuals. Note, though, that I am thereby choosing simplicity over precision. I trust that the reader will bear in mind that problems with the limit assumption can be resolved by employing the more sophisticated STANDARD-1 instead.

So far, we have only looked at counterfactuals with false antecedents. Let us now briefly look at a less common, but equally important case for our concerns in this chapter: counterfactuals with true antecedents. Consider the counterfactual "If I had a dog, I would go for a walk every day", and suppose I actually have a dog. How do we determine the truth conditions of such a counterfactual? As before: move to the closest antecedent-worlds and determine whether the consequent is true in them. Which ones are the closest antecedent-worlds, though? Here, people diverge.

Most people accept what is called "centering": the idea that every world is closest to itself and no other world is as close. Given centering, the closest antecedent-worlds for counterfactuals with true antecedents will always be only the actual world itself. Thus, to find out whether it is true that I would go for a walk every day if I had a dog, all we need to check is whether I actually go for a walk every day. If I do, the counterfactual is true. If I don't, it's false.

But not everyone endorses this part of the semantics. Nozick (1981: 176), for instance, rejects it, and holds that the sphere of the closest antecedent worlds will contain a few non-actual worlds, even in the case of counterfactuals with true antecedents. Hence, to determine whether or not I would go for a walk every day, if I had a dog, it will not do to determine whether or not I go for a walk every day in the actual world. It will also have to be true that I go for a walk every day in the other closest, non-actual worlds in which I have a dog. The counterfactual is true only if I go for a walk every day in those worlds as well.

Whether or not centering is a good idea will not – and need not – be settled here, because it is irrelevant to the arguments that are going to be developed in what follows. For simplicity, I will assume that it holds. Whenever I anticipate that issues about centering could be thought relevant to one of my arguments, I'll provide a brief commentary on why the argument works either way.

This very brief introduction into the semantics of counterfactuals will do. A counterfactual (with a possible antecedent) is true if and only if the consequent is true in the closest antecedent worlds. What we need to do now is apply this to the simple conditional analysis.

The simple conditional analysis states that an agent has an ability to φ if and only if the agent would φ, if she intended to φ. What this translates into, on the standard semantics, is that an agent has an ability if and only if the agent φ's in the closest possible worlds in which the agent intends to φ. Thus, to find out whether an agent has an ability, we need to move to the closest worlds in which the agent intends to φ and check whether she effectively φ's in those worlds. If she does, she has the ability. If she doesn't, she lacks it.

Quite intuitive, at first glance. As we'll see in the next sections, though, the view is not very plausible at all, because it runs into problems with so-called "masks" (2.2) as well as cases of impeded intentions (2.3), and it fails to account for the distinction between general and specific abilities (2.4), degrees and the corresponding kind of context sensitivity of ability ascriptions (2.5), as well as for non-agentive abilities (2.6).

2.2 Extensional inadequacy I: Masks

In this and the next section, we will see that the conditional analysis is extensionally inadequate; the view invites counterexamples which show that the conditional analysis is neither necessary nor sufficient for an agent to have an ability. In this section, we will focus on the first type of counterexamples – counterexamples which show that the conditional analysis is not necessary.

As I'll argue, then, it may well be that some agent has an ability to φ, and yet it may be false that the agent would φ if she intended to φ. How? Simply thus: something may interfere with the exercise of the ability. Call an ability which is had but whose exercise would be interfered with a *masked* ability, and call the interfering factor a *mask* (Johnston 1992). Here are some everyday examples of masked abilities. My ability to swim may be masked by a cramp. My ability to sing a certain song may be masked by my hoarseness or my shyness or my forgetfulness of the lyrics. My ability to hit the bull's eye may be masked by my sudden dizziness.

In all of those cases, the counterfactual is false: it is not true that I would swim or sing or hit the bull's eye if I intended to do so, because the closest intention worlds will be worlds in which the mask will be present. In the closest worlds in which I form the intention, I thus fail. Yet I do have the relevant abilities. Hence, masking cases show that the truth of the counterfactual is not necessary.

There is an even simpler way to come up with masking cases: all we need to do is cite cases in which an agent who has an ability to φ *actually* intends to φ and fails. Whenever an agent has an ability to φ, but fails to realize her intention to φ, we have a case of masking: something came in the way. Again, the counterfactual is false in such a case: it is not the case that the agent would φ, if she intended to, because she *did* intend to φ and *didn't* φ.

This is one of the places where centering may be thought to play a role in the argument. If centering is false, then the closest worlds contain not just the actual world, but also other worlds in which the agent intends to φ. And couldn't it be that what masks the ability in the actual world is not present in those worlds? The smaller and more random the feature is which interferes with the exercise of the ability in the actual world, the more plausible it is that the same interference is not present in the other closest worlds, after all. Thus, one may think, it does not do to cite an actual intention and failure to falsify the counterfactual, if one rejects centering.

As far as I can see, this line of thinking is misled, though. Even *if* the closest worlds contain some in which the interfering factor is not present, this will not help to rescue the truth of the counterfactual. To see that, let's go through the truth conditions very carefully. First, let us assume that the limit assumption is true, and a counterfactual is therefore true if and only if all of the closest antecedent worlds are consequent worlds. In that case, the conditional analysis postulates that the agent φ's in all of the closest worlds in which she intends to φ. Let's now assume that centering is false. The closest intention worlds will now contain some non-actual worlds that are still very close. Let's also assume, for the sake of the argument, that the agent does in fact φ in some of those closest worlds.

The important point is that the counterfactual will still be false in that case. For as long as the actual world is *among* the closest intention worlds, a failure to φ in the actual world will suffice to falsify the counterfactual. Rejecting centering is to reject that the actual world is the *sole* member of the set of closest worlds. It is not to reject that the actual world is *among* the closest worlds. Whether or not we accept centering: the actual world will always be among the closest worlds to itself. Hence, the argument runs smoothly, no matter whether or not centering is accepted or not.

What if we also reject the limit assumption? Then we will have to work with the slightly more complicated truth condition for the counterfactual to be true. But even then, the counterfactual will be false, if the actual world is an antecedent, but not a consequent world.

Suppose the limit assumption is false. Then a counterfactual is true if and only if there is some antecedent world in which the consequent is true that is closer than any antecedent world in which the consequent is false. Thus spelled out, the simple conditional analysis postulates that there be some intention world in which the agent φ's that is closer than any intention world in which the agent does not ϕ. Let's again reject centering. The sphere of the closest intention worlds will be slightly larger than the sphere containing only the actual world. And let's assume, for the sake of argument, that the agent does in fact succeed to ϕ in some of the closest intention worlds.

Even so, the counterfactual will be false, given that the agent intends to ϕ and fails in the actual world. That is because it is not true, in that case, that some world in which the agent intends to ϕ and ϕ's is closer than any world in which the agent intends to ϕ and fails. There may be *closest* worlds in which the agent intends to ϕ and succeeds, but none of them will be *closer* to the actual world than the actual world itself. They will be just as close, but not closer. No matter how we position ourselves regarding the various ways of analyzing counterfactuals, an actual intention and failure suffices to falsify the counterfactual figuring in the conditional analysis.

Masks provide a simple lesson, then. The counterfactual is too strong to be a necessary condition for an agent to have an ability. Abilities can be had, and yet it may be true that an intention would fail to be realized. Actual intentions *do* quite often fail to be realized. And if no actual intention is formed, the world may be such that it *would* fail to be realized. This is all we need to show that the simple conditional analysis fails to state a necessary condition for an agent to have an ability.

Can the conditional analysis be ameliorated? At first sight, it may seem like a good idea to add specifications into the antecedent to exclude masking cases. Aune suggests that the relevant conditional in the case of abilities is not "I will if I choose", but rather "if I want to lift it, try to lift it, *and nothing interferes with my attempt*, I will certainly succeed" (Aune 1963: 338, my emphasis).[5] On closer inspection, though, this strategy proves problematic.

To see that, let's take a step back from abilities and look into the recent literature on dispositions. For readers familiar with that literature, masks and their impact on conditional analyses are not a dark horse. Conditional analyses have been very prominent not just in connection with abilities, but also, and in fact foremost, in connection with dispositions.

According to what we can call *the simple conditional analysis of dispositions*, something has a disposition to exhibit some behavior if and only if it would exhibit that behavior if it were to undergo an appropriate stimulus (Ryle 1949; Goodman 1954; Quine 1960). Sugar is soluble, on this view, if and only if it would dissolve if it were placed in liquid. A poison is toxic if and only if it

---

[5] We also find the formulation: "if Jones were to try to do X under absolutely noninterfering, or nonfrustrating, conditions, he would be successful" (ibid.: 401).

would cause harm if it were ingested. A person is irascible if and only if she would freak out if she were provoked.

The analogy between the conditional analysis of abilities and the conditional analysis of dispositions should be obvious. In both cases, a counterfactual tie is postulated between some trigger and the manifestation of the ability and the disposition, respectively. In view of this similarity, it does not come as a surprise that cases of masking have been raised as a problem against the conditional analysis of dispositions as well. A poison may be ingested together with an antidote, in which case it is not true that the poison would cause harm if it were ingested (Bird 1998).[6] An irascible person may be provoked, but her wife may be around to soothe her. In that case, she would not freak out if she were provoked. And so forth.

Masking is a very prominent objection against the simple conditional analysis of dispositions and is extensively discussed in that connection. And of course, we find an analogue of Aune's proposed amelioration of the counterfactual in connection with dispositions, too. Most famously, the analogous suggestion is made by Lewis when he writes that

> we might offhand define a poison as a substance that is disposed to cause death if ingested. But that is rough (...) [W]e should really say 'if ingested without its antidote'. (Lewis 1997: 145)

The suggestion is clear. Just as Aune suggests for abilities, the idea is to retain the general scheme of the analysis, but to specify the antecedent in such a way that masks are explicitly ruled out. A number of authors (i.e. Choi 2006, 2008; Gundersen 2002; Mumford 1998, ch. 4) have followed Lewis's lead. While the accounts differ in their exact formulation, they all state essentially the same idea: x has a certain disposition to φ if and only if x would φ if a stimulus occurred and conditions C obtained, where conditions C will have to be understood to exclude masks. Following Whittle (2010), we can call this the *specifier approach*.

There is definitely something to this approach. It certainly calls attention to a certain discomfort I believe most people have when they first encounter the objection of masks. Somehow it seems clear that when we say things like "to be toxic is to cause harm, if ingested", we implicitly intend "if ingested" to be understood as something along the lines of "if ingested *under the right conditions*". Analogously for the conditional in "to have an ability to swim is to swim if one tried to swim". Of course, what is meant is "if one tried to swim *under the right*

---

6 Bird calls masks in general "antidotes".

conditions". And if that's what is meant then coming up with masks is just cheating; it exploits the fact that we have not made that additional condition explicit right away. Once it is made explicit, however, those apparent counterexamples turn out to be simply irrelevant for the truth of the conditional.

Legitimate as this intuition seems, however, there are two problems with the specifier approach. The first is that it is far from clear whether C – the good conditions – can be spelled out in a non-trivial way at all. That is because for each disposition and ability, respectively, there is an indefinite variety of potential masks:

> The kazoo's disposition to buzz when blown through (...) can be masked by stuffing the kazoo full of paper. It can also be masked by coating the kazoo's vibrating reed with wax; or by putting one's finger over the end of the kazoo; or by submerging the end of the kazoo in honey; or .... Given any finite list of potential maskers (...), it seems that a masking situation not on that list could always be imagined. (Fara 2005: 51)

Fara's point is clear. No specification of the circumstances could ever be complete when it comes to the exclusion of the crucial counterexamples. It thus seems that the only way of specifying C would have to be the specification of a feature that all of the potential counterexamples have in common. Given the diversity of the examples, however, it is hard to think of such a feature.

Indeed, as Fara rightly points out, "the only relevant property that cases of masking have in common with each other is that they *are* cases of masking" (ibid.). And this shared property helps little when it comes to the specification of C. For surely, we should not say that something has a disposition to φ if and only if it would φ in the absence of masks. So formulated, the conditional states a triviality; it states that something has a disposition to φ if and only if it φ's as long as nothing prevents it from doing so. Obviously, on this account, every object has every disposition whatsoever (Bird 1998: 231).

Abilities do not differ from dispositions when it comes to the diversity of potential masks: my ability to swim can be masked by my legs being bound together, by some weight on my feet, by panic, by a cramp, by too many people leaving too little space... But those masks do not seem to have anything in common besides the fact that they prevent my ability to swim from being exercised. Hence, the problem of the plentitude and diversity of potential counterexamples arises for abilities as well; it will either turn out impossible to specify C or we are once more heading towards the triviality problem.

Aune's suggestion falls prey to the second horn of the dilemma: stating that to have an ability is for it to be true that the agent would φ, if she intended to and nothing interfered, does indeed seem trivial. For what else is an interference than simply a feature of the situation that results in a non-occurrence of whatever actions is intended from being performed?

The second problem with the specifier approach is related, but different. For even *if* the circumstances which pose an interference could be specified, the specifier view will run into problems. To see that, suppose you and I both have the ability to hit the bull's eye. We both hit it in seven out of ten cases. In three out of ten cases, the ability is masked. But now suppose those three cases are not the same ones in your case and in mine. Whenever you fail, I succeed. Whenever I fail, you succeed. We both have the ability to hit the bull's eye, but that ability is masked in completely different conditions.

This case poses a problem for the specifier approach, because the specifier approach does not provide any means to specify our *shared* ability to hit the bull's eye. Your ability will be the ability to hit the bull's eye in conditions in which the set of impediments A does not obtain, whereas my ability will be the ability to hit the bull's eye in conditions in which a different set of impediments B does not obtain. Even if the impediments could be specified in a non-trivial way, then, the specifier view fails to account for the fact that one and the same ability may be masked by completely different factors in different agents (cf. Manley and Wasserman 2008: 75n20).

We can conclude that masks are problematic for the conditional analysis: there is good reason to doubt that there is a non-trivial counterfactual, the truth of which will turn out to be necessary for the having of an ability.

2.3 Extensional inadequacy II: Impeded intentions

Masks show that the conditional analysis does not state a necessary condition for the having of an ability. Let me now argue that it also fails to state a sufficient condition. To see the problem, let's recall that an agent has the ability to φ, according to the simple conditional analysis, if and only if she would φ, if she intended to φ. Why is that condition not sufficient? The reason is very easily stated: it may well be true that an agent would perform a certain action *if* she intended to perform it, but be unable to form the intention to begin with. In that case, the counterfactual is true, while the ability is lacking (Moore 1912, ch. 1 & 7; Chisholm 1966, 1976; Lehrer 1968; van Inwagen 1983: 119; Whittle 2010).

Illustrations are easy to come by. Comatose patients (van Inwagen 1983: 119) lack the ability to raise their arms. Yet, the conditional analysis yields that they have the ability. That is because the counterfactual turns out true for the coma patient: if she intended to raise her arm, she would. To see that, let's frame the issue in the standard semantics for counterfactuals. To find out whether the coma patient has the ability to raise her arm, we have to move to the closest possible worlds in which she intends to raise it, according to the conditional analysis. But since the coma prevents the agent from forming an intention to raise her arm in the first place, we have to move *beyond* the coma worlds in order to reach the closest possible intention worlds. Since the agent is not in a coma in those worlds, however, there is no reason to assume that she would not succeed in raising her arm in those worlds. Hence, the counterfactual turns out true, while the agent obviously lacks the ability.

In case you wonder whether it makes a difference which motivational state figures in the counterfactual, let me point out that it does not. The coma case is a counterexample to any version of the conditional analysis. A coma prevents an agent from forming any motivational state of raising her hand whatsoever – coma patients can neither try to raise their hands, nor can they intend, choose, or want to raise them. No matter what motivational state we insert, the analysis fails.

Can we think of further cases with such an impact? I think we can. Just consider a brainwashed follower of a cult.[7] Can the person leave the cult? No. Being as brainwashed as she is, she cannot. But would she leave the cult, if she formed the motivation to leave? Presumably yes. What the brainwash plausibly does is prevent the agent from forming the motivation to leave to begin with. Hence, the closest motivation worlds will be worlds in which the agent is not brainwashed. And hence, there is nothing to prevent the agent from leaving the cult in those worlds.

Again, it seems that the counterexample works against all versions of the conditional analysis alike. Brainwash prevents the agent from forming any motivation to leave whatsoever. Intending, choosing, wanting, and trying to leave are all beyond the range of things a thoroughly brainwashed person can do. Hence, the example seems to refute the conditional analysis as such, and not just a particular version of it.

What all this shows is that the conditional analysis runs into deep structural problems with a certain class of cases. Whenever an agent cannot φ in virtue of an impairment that prevents her from forming the intention (or any kind of mo-

---

[7] An example that is often cited is that of a phobic, who cannot touch spiders because she cannot form the intention or make the attempt to begin with (Lehrer 1968). I don't find that example particularly well chosen, because it is not entirely clear whether the phobic is really impaired on the motivational level. Furthermore, the phobia case invites discussions about potential variances in plausibility between the various versions of the conditional analysis with different motivational states in the antecedent. For those reasons, I prefer to talk about coma patients and brainwashed agents instead.

tivation) to φ to begin with, the counterfactual will come out true, while the agent obviously lacks the ability to φ. Let's call this the *problem of the impeded intention*.

It is important to note that the problem of the impeded intention does not hinge on the claim that there is *no* good sense of "ability" in which the coma patient *does* in fact have the ability to raise her arm (and likewise for the brainwash case). Presumably, there is such a sense. The ability to raise one's arm may not be the best example to see that. But consider abilities which actually require a lot of expertise. Suppose the coma patient is a world-famous juggler. There is, I take it, good sense to be made of the claim that the coma patient still has the ability to juggle.

That this claim can be made perfect sense of does not attenuate the problem of the impeded intention, however. There may be a good sense in which the coma patient can juggle. Yet, there is an equally good sense in which she cannot juggle. The reason for the lack of that ability is obviously that the coma patient is in a coma. Obviously, a coma is an impediment to juggling, and a very severe one at that. Hence, a comatose agent is unable to juggle (in a very sense of "ability"). I don't think there is much room for disagreement here. The problem with the conditional analysis is that it fails to account for that very good sense in which the coma patient lacks various abilities.

The problem of the impeded intention is a very deep-running problem, then, because it shows that the whole structure of the counterfactual is ill-suited when it comes to analyzing abilities. The ability to φ depends, in part, on the ability to form the intention to φ. And so, the mere fact that there is a counterfactual connection between the intention and the agent's performance is, by its very nature, insufficient to show that the agent has the ability. When the agent cannot form the intention to φ to begin with, she also cannot φ. The simple conditional analysis seems to exhibit a deep structural flaw at this point.

There is a natural rejoinder for the proponent of the simple conditional analysis. If the problem of the impeded intention is that the agents in the examples cannot form the intention in the first place, why not simply add a second condition? Why not simply postulate that the agent also has to be able to form the intention in the first place? This further requirement, it can be argued, is already implicit in the conditional analysis, and once pressed, the proponent of the conditional analysis may simply state it as an additional condition of her view. According to the full account, then, there has to be a counterfactual connection between the agent's intentions and performances on the one hand, but the agent also has to be able to form the intention in the first place.

As we will see in connection to the account of abilities that I will lay out in chapter 4, this strategy is actually a very good idea. The ability to φ does indeed require the agent also to have the ability to form the intention to φ. The problem is just that it cannot be spelled out with the toolbox available to the proponent of the conditional analysis as it stands. For what is it to be able to form the intention to φ? It is to have yet another ability. But surely, it will not do to simply reapply the conditional analysis to that ability all over: it is not a very promising move to suggest that the ability to intend to φ is had if and only if the agent would form the intention to φ, if she intended to form the intention to φ.[8] The analysis becomes either circular at this point or we slip into a regress. Either way, we will not have made much progress.

The problem remains. As the coma case and the brainwash case show, it may well be that an agent would φ if she intended to φ, and yet she may be utterly unable to φ. We can conclude that the simple conditional analysis, besides failing to state a necessary condition for an agent to have an ability, also fails to state a sufficient condition.

2.4 The problem with general and specific abilities

In the last two sections, we saw that the simple conditional analysis runs into extensional problems. In this and the next two sections, I'll show that the simple conditional analysis also fails to account for the other three adequacy conditions within a comprehensive view of abilities.

Let's start with the condition that any comprehensive view of abilities will have to elucidate the relation between general and specific abilities. As I will argue, the simple conditional analysis fails to account for both general and specific abilities individually. Consequently, the relation between the two remains in the dark as well.

That the simple conditional analysis fails to provide an account of general abilities is quite easy to see. Take the general ability to do a handstand. Obviously, you can have that ability, even if you are presently dizzy or on a shaky boat, say. That's just what makes the ability general: whether or not you have them does not hinge on the temporary features of your current situation; temporary unfavorable circumstances do not deprive you of your general abilities. The specific ability to do a handstand, in contrast, *does* hinge on such features. You cannot do a handstand here and now when you are here and now dizzy or on a shaky boat.

---

[8] Moore (1912, ch. 7) makes this suggestion: having the ability to choose, he notes, is for it to be the case that the agent would choose, if she chose to choose.

It should be quite obvious that the simple conditional analysis fails miserably as an account of general abilities. A counterfactual is true if and only if the closest worlds in which the antecedent is true are also worlds in which the consequent is true. To find out whether a counterfactual holds, that is, we move away from the actual world only far enough to make the antecedent true. To find out whether someone would do a handstand if she intended to, say, we move to the very closest worlds in which the agent intends to do a handstand. As much as possible remains as it is in actuality in those worlds. Hence, a dizzy agent is dizzy in those worlds, and an agent on a shaky boat is still on a shaky boat in those worlds.

Will the agent do a handstand in those intention worlds? Of course not. The counterfactual will come out false for dizzy agents and agents on shaky boats, then. But we said that the agent can have the ability to do a handstand despite her dizziness and unfavorable location, respectively. Hence, the counterfactual does not do justice to abilities that are general in that way.

Quite generally put, the problem with an account of general abilities in terms of the conditional analysis is that a general ability abstracts away from very many features of an actual situation, whereas the semantics of a counterfactual factors all of the features of the actual situations in when determining the closest antecedent worlds. Hence, the tension between general abilities and a counterfactual account of such abilities runs rather deep.

How about specific abilities? At first one may think that the conditional analysis is rather well suited as an account of abilities of that kind. Especially in view of the failures it exhibits with respect to general abilities, this may seem quite plausible at first. A specific ability is an ability to perform an act in a given situation. It thus seems like a good idea to look at what the agent would do in an intention situation that is as similar as possible to that very situation. Hence, it seems as though specific abilities do not conflict with the semantics of the counterfactual in the destructive way as general abilities. When I am dizzy or on a shaky boat, I cannot do a handstand here and now, with those impediments in place. When it comes to specific abilities, it therefore seems to be a good feature of the semantics of counterfactuals that we look for intention worlds that resemble the actual world as closely as possible with respect to as many features as possible – including the impediments to the exercise of the ability that are actually in place.

The plausibility of this line of thought hinges on a topic we touched on earlier: the maskability of abilities. To see that, note that the reason I cited for the inappropriateness of the conditional analysis for general abilities can be formulated in terms of their maskability. General abilities can obviously be masked. The actual world can contain impediments to the exercise of the ability without thereby depriving us of the ability altogether. Dizziness and being located on a shaky boat are masks for one's general ability to do a handstand, for example.

The important point is that it is *in virtue of the maskability* of general abilities that that such abilities cannot be accounted for in terms of a counterfactual. The general ability is there despite the masks. Yet, the mask prevents the exercise of the ability. But since the worlds relevant to the truth of the counterfactual are the closest intention worlds of the agent, the mask will usually be present in such worlds. Hence, the counterfactual will turn out false whenever the actual world contains a mask for an agent's general ability.

We can conclude from this that the simple conditional analysis runs into problems with any maskable ability whatsoever. So the question we need to answer in order to find out whether or not the conditional analysis is suitable as an account at least of specific abilities is whether or not specific abilities can be masked or not. If they cannot, the conditional analysis will very plausibly state at least a necessary condition for the having of such abilities. If they can be masked, in contrast, then the simple conditional analysis is no more plausible as an account of them than of general abilities.

The question whether or not specific abilities are maskable is not easily answered; philosophers seem to have very different ideas about this question (e. g. Fara 2008; Vihvelin 2004; Clarke 2009). What is indisputable, though, is that unless virtually *every* fact, no matter how random or small, of the actual situation matters for whether or not some agent has the specific ability to perform some act in that situation, masks will be perfectly possible. Unless virtually every fact counts, there can always be a feature of the actual world that would come in the way of the exercise of an ability.

Does every fact count for an agent's specific abilities? Let's test your intuitions. Suppose you intend to hit the bull's eye and you fail, even though you usually hit it and the situation was perfectly favorable to your hitting it. Does your failure in that situation show that you did not have the specific ability to hit your target? Only when you answer this in the affirmative should you subscribe to the idea that specific abilities cannot be masked. If, in on the other hand, you believe that there is at least one good sense of "specific ability" according to which you did have the specific ability in that situation, albeit one that you failed to exercise, you should instead commit to the idea that even specific abilities can be masked.

If the latter is true – if some specific abilities can, in fact, be masked – then the simple conditional analysis will not yield the proper account of specific abilities, either. There can always be some influence present in virtue of which it is not true that the agent would perform some action, if she intended to perform it.

Maskable specific abilities do not differ from general abilities at all in this respect.

My own view on specific abilities, upon which I will elaborate in chapter 4.5, is that there is indeed a good sense of "specific ability" according to which specific abilities can be masked. I take it to be reasonable that there are more or less specific abilities among the ones we rightly think of as specific. In fact, I believe that the most fruitful picture of the distinction between general and specific abilities is a more or less scalar one, according to which there are more or less general and specific abilities, with maximally specific ones on one end of the scale and maximally general ones on the other. They differ with respect to the set of facts that matters for whether or not the agent has the ability. Thus, I believe that even fairly specific abilities can be masked, while the most specific ones are essentially unmaskable (→ 4.5d).

If this is correct, then the conditional analysis will at best state a necessary condition only of the most specific abilities we can think of – abilities which are such that any failed attempt shows that the agent did not have the ability in the first place. Slightly less specific abilities, abilities for which lots of facts about the actual situation matter, but which still allow for masks, in contrast, will be as badly accommodated by the counterfactual as any general ability.

Moreover, we should be very clear on one thing. Despite the fact that the simple conditional analysis may actually state a necessary condition for some very, very specific abilities, it is nevertheless inappropriate as an *analysis* of abilities of that kind. The reason for that has to do with the problem of the impeded intention once more. The problem of the impeded intention is not limited to general or even fairly general abilities at all. Quite to the contrary. While there may be some sense in which even the coma patient retains the very *general* ability to raise her arm, there is obviously no sense in which the coma patient has the *specific* ability to raise her arm in her particular situation, coma and all. Hence, the problem of the impeded intention seems even more pressing with respect to specific abilities, and *a forteriori* particularly pressing with respect to the most specific ones among the already specific abilities.

2.5 The problem with degrees and context sensitivity

In the last section, we saw that the simple conditional analysis fails to provide an account of general and specific abilities. In this section, I'll argue that the view also fails to account for the third adequacy condition: it fails to offer an account of degrees and the corresponding kind of context sensitivity that attaches to ability ascriptions.

Abilities, we said in chapter 1, come in degrees. Glenn Gould has the ability to play the piano to a higher degree than I do; my dog has the ability to smell to a higher degree than most of us; my grandmother sews better, my cousin drives better, my father draws better than I do. One meteorologist may be better at predicting the weather than another. Two archers may be varyingly good at hitting the target. We practice things to become better at them. Abilities can be improved. We admire people who have outstanding abilities in some field or another. In a nutshell: degrees of abilities are phenomena with which we are all most familiar.

Looking more closely at degrees of abilities, we saw that the degree of an agent's ability can be influenced on two dimensions. The degree of an ability depends, first, on the quality of the agent's performances, and secondly on the range of circumstances across which the agent is able to deliver those performances. We called those two dimensions *achievement* and *reliability*, respectively.

We also said that the gradable nature of abilities results in a specific kind of context sensitivity of ability ascriptions. The degree of an agent's ability to φ required for the agent to count as having the ability to φ *simpliciter* varies across contexts. A darts player who misses 99% of the time, but who manages to hit the bull's eye every 100th time does not usually count as having the ability to hit it. There may be contexts in which we count her as having the ability, though. She has what it takes; it is within the range of what she can do. The context, we said, determines some threshold along the scale of degrees of abilities above which an agent counts as having the ability *simpliciter*. Once we have degrees, then, we also have the mechanism for the corresponding context sensitivity.

The problem with the conditional analysis is that it is structurally ill-suited to yield an account of degrees of abilities. That is not to say that it fails to account for all variances in the degrees of two agents' abilities whatsoever. The conditional analysis can account for the fact that my partner is a better driver than I am, for instance, because (ignoring the problem of masks) it will rightly yield that while my partner can drive, I cannot. If he intended to drive, he would succeed. If I intended to drive, I would fail – I usually stall the engine right away. Given that anyone who has an ability at all has that ability to a higher degree than anyone who completely lacks it, my partner will come out more able than I.

But apparently, that does not give us the full range of degrees we are striving for. It does not, for instance, yield that my partner is a better driver than his friend. The problem with my partner's friend is not that she cannot drive at all. The problem is rather that she is not particularly good at it. This difference between my partner and his friend cannot be captured in terms of the counterfactual, though. Both would drive if they intended to. Yet, one of them would drive perfectly fine, whereas the other one of them would drive crazily.

Now, you may want to respond that this is not as much of a problem as it seems at first sight. Here is a suggestion for how the proponent of the conditional analysis may account for the difference between my partner and his friend. In a first step, let's attach values to the quality of an agent's performance. My partner gets a 10 and his friend gets a 1, say. And now let's specify the quality of the agent's performance in the consequent. Then we get the following two conditionals: if my partner intended to drive, he would drive with quality 10. If his friend intended to drive, she would drive with quality 1. We can now say that the degree of an agent's ability is higher the higher the value attached to the agent's performance in the consequent.

This seems like a reasonable suggestion, but unfortunately it does not take us very far, either. First of all, the problem of masks re-emerges with new force. Suppose my partner is drunk. Then he, too, would drive crazily, if he intended to drive. In such circumstances, the simple conditional analysis, as we have just constructed it, yields that he is not good at driving at all. Again, the problem seems to be that the counterfactual ties agents' abilities way too closely to the circumstances they are actually in. This is a problem for the account of degrees to the same extent as it is a problem for the simple conditional analysis quite generally.

The problem I have just pointed out does not have primarily to do with *degrees*, though – in fact, it is just the problem of masks in a slightly different guise. (What it shows, though, is that the problem of masks is getting more and more complicated the closer one looks at it.) What really strikes me as a problem specifically in connection with degrees is something else: the problem is that conditionalists cannot account for the dimension of reliability at all.

How are we to measure the respective degrees of two meteorologists' abilities to predict the weather correctly, for example? Let us call our two meteorologists Fred and Ted. Fred has developed an algorithm, by means of which he can predict 99 percent of all weather constellations correctly. Ted has developed a faulty algorithm – it goes wrong for 45 percent of all weather constellations. Clearly, Fred is better at predicting the weather. Nevertheless, the counterfactual will often come out true for both of them alike. For suppose we are in a situation in which the weather is be predicted correctly by both meteorologists alike. Then the counterfactual will come out true for both of them. Who is the better meteorologist? According to the conditional analysis, they are on a par.

The problem that degrees of abilities pose for the conditional analysis is thus again that the counterfactual is too closely tied to the agent's actual situation. To account for degrees and specifically for the dimension of reliability, we need a means to evaluate the agent's performances across various kinds of circumstances. The counterfactual, however, only ever evaluates the agent's performance in *one* set of circumstances: the circumstances in which the agent intends to ϕ and which are otherwise as similar as possible to the agent's *actual* circumstances.

2.6 The problem with non-agentive abilities

There is one adequacy condition left to discuss. Any comprehensive view of abilities will have to account for agentive as well as non-agentive abilities. It should be obvious that the simple conditional analysis does not look too well-equipped with respect to that condition either. The simple conditional analysis, with its focus on the counterfactual tie between intentions on the one hand and performances on the other, is tailor-made for *agentive* abilities – abilities to perform paradigmatic instances of intentional actions. In cases of that kind, it is quite appealing to think that the having of such an ability has something to do with the tie between the agent's intention to do ϕ and her effectively doing ϕ. To have the ability to jump, it has to be true that you would jump, if you intended to jump. To have the ability to sing, it has to be true that you would sing, if you intended to sing. As I argued, the simple conditional analysis fails even here, but at least it looks very plausible at first.

In the case of non-agentive abilities, however, the simple conditional analysis looks highly suspicious even at the outset. Non-agentive abilities are abilities to be engaged in certain behaviors that are not actions. It is therefore crucial for non-agentive abilities that they can (or can only) be exercised without a foregoing intention on the agent's part. A view which determines whether or not an agent has an ability by looking at the tie between intentions and performances thus seems inadequate right from the beginning.[9]

Now, perhaps the proponent of the simple conditional analysis need not admit defeat right away. In defense of her view, she might say: Look, it may seem odd to analyze abilities like the ability to digest or to produce saliva in terms of a counterfactual tie between intentions and performances. But in fact the analysis yields the right results. That is because of course I would digest, if I intended to. All I would have to do is eat something. And of course I would produce saliva, if I intended to. All I would have to do is bite my tongue.

---

[9] Note that, again, nothing hinges on the particular motivational state that figures in the analysis. I will not comment on this any more as I go along.

Thus, the counterfactual comes out true for those kinds of abilities. That is all that counts.[10]

There are at least two reasons why this strategy is unsuccessful. First, it yields an ugly disconnection between the conditions for *having* an ability and the conditions for *exercising* an ability.[11] It seems very natural for the proponent of the conditional analysis to pair her claim that an ability to ϕ is *had* if and only if the agent would ϕ, if she intended to, with the further claim that an ability is *exercised* if and only if the agent's ϕ-ing results (non-deviantly) from an intention to ϕ. In the case of agentive abilities, this seems very plausible. If I hit the bull's eye upon intending to hit it, I have exercised my ability to hit it. If I hit it by accident, though I actually intended to hit the outer ring, say, I take it that I have not exercised the ability to hit the bull's eye. My hitting the bull's eye is a mere accident in that case. The simple conditional analysis has a good story to tell about this: an ability is exercised if and only if the action follows an intention to perform it.

In the case of non-agentive abilities, things look rather different. Obviously, not every instance of digesting is a result of intending to digest. Indeed, the intended instances of digestion are very rare; usually, the digestion occurs without any foregoing intention on the agent's part. Yet, one will surely want to say that the unintended instances of digesting are, in fact, exercises of the agent's ability to digest. But here, the proponent of the conditional analysis does not have a good story; and whatever she says here will either result in a disanalogy with the agentive case or – if both cases are treated along the same lines – with the consequence that my hitting the bull's eye by accident counts as an exercise of my ability to hit the bull's eye as well. That does not seem auspicious.

Even if the proponent of the simple conditional analysis is willing to go for one of those options, however, there is a second and even stronger reason why non-agentive abilities cannot be accounted for by in terms of a counterfactual link between intentions and performances: there are some non-agentive abilities to ϕ which are *never* preceded by an intention to perform ϕ. Just think of the examples I cited in chapter 2.6 – the ability to unintentionally read street signs, for instance. Or the ability to hear what someone says without listening. With respect to those examples, the limitations of the simple conditional analysis jump right at us. Clearly, it is not the case that I would read street signs unintentionally, if I intended to. Once I intend to unintentionally read street signs, I am bound to fail.

---

10 Erasmus Mayr (personal conversation) suggested this strategy to me, without, I think, endorsing it.
11 For the same point in connection with dispositions, see Vetter (ms.).

Thus, the simple conditional analysis yields that I lack the ability to unintentionally read street signs. This is the wrong result.

What seems rather strange even in the case of digesting and producing saliva is therefore obviously wrong when it comes to other non-agentive abilities. We can conclude that abilities of the non-agentive kind are not properly accounted for by the simple conditional analysis. The modal tie between intentions and performances, while very plausible for agentive abilities, is obviously ill-suited to account for abilities of the non-agentive variety.

As we will see in chapter 4, my own view bears some similarities to the simple conditional analysis in that it, too, analyzes agentive abilities in terms of a modal tie between intentions and performances. I want to emphasize very clearly, however, that it does not stop there. For as I'll show in chapter 5, the view can be neatly extended to cover non-agentive abilities as well. However, the extension I suggest for my own view could, in principle, just as well be integrated into a framework operating with counterfactual conditionals. I'll briefly get back to all this in chapter 6.2, when I discuss the sophisticated conditional analysis I have been talking about earlier.

Since we are so far dealing with the conditional analysis as it is, however, let us just note that, as it stands, the view fails to provide the tools that would be needed to account not just for agentive, but also for non-agentive abilities. Together with the problems we noted in the last four sections, we therefore get the result that the simple conditional analysis runs into trouble with all of the adequacy conditions we have identified for a comprehensive view of abilities. Hence, the simple conditional analysis is certainly not the way to go.

2.7 Upshot

Let me once again walk you through the main insights of this chapter. In this chapter, we looked at what I called the *simple conditional analysis*. I used that notion as an umbrella term for any view which analyzes abilities in terms of a counterfactual tie between some motivational state to φ and the agent's effectively φ-ing. For plasticity, I focused on one specific version of the view, namely that an agent has an ability if and only if she would φ, if she intended to φ.

Five problems were laid out for the simple conditional analysis, all of which consisted in failures to do justice to one of the adequacy conditions for a comprehensive view of abilities.

First, the simple conditional analysis fails to state a necessary condition, which was shown by pointing to cases of masking (→ 2.2). In masking cases, there is some impediment which would prevent the agent from exercising her ability upon intending to do so. A cramp, for instance, masks the ability to swim. I argued that masks are a rather severe problem for the simple conditional analysis, because they show that the ability can be had, while the counterfactual turns out false. I also argued that the so-called specifier approach – an amelioration of the view that tries to exclude masks by specifying the antecedent – is much harder to spell out than one might think.

Secondly, the simple conditional analysis also fails to spell out a sufficient condition. This was established by showing that it may well be true that the agent would φ, if she intended to φ, and yet the agent may be unable to φ, because she may be unable to form the intention to φ to begin with. I called this "the problem of the impeded intention" (→ 2.3). Masks and impeded intentions show that the simple conditional analysis is extensionally inadequate.

Thirdly, the simple conditional analysis runs into problems with general and specific abilities (→ 2.4). General abilities elude the analysis completely, because it is a characteristic feature of general abilities that they can be had, even if the agent is currently in unfavorable circumstances. Since the counterfactual is always evaluated at the closest intention worlds, the counterfactual will come out false in such cases. That is a problem.

The view works better with respect to specific abilities, although a lot hinges on the highly controversial question of whether or not specific abilities can be masked. Even if they cannot be masked, however, the simple conditional analysis will at best state a necessary condition for the having of such an ability, since the problem of the impeded intention is particularly powerful when it comes to specific abilities.

Fourthly, the simple conditional analysis fails to account for degrees of abilities and the corresponding kind of context sensitivity that attaches to ability statements (→ 2.5). The degree of an ability to φ has to do, in part, with the range of possible circumstances across which the agent is successful in φ-ing. Since the counterfactual only ever evaluates the closest circumstances in which the agent intends to φ, however, the simple conditional analysis does not provide the tools to account for this fact. Consequently, the corresponding kind of context sensitivity of ability statements – one that has to do with a variable threshold on the scale of degrees – remains in the dark as well.

Finally, the simple conditional analysis fails to account for non-agentive abilities (→ 2.6). The tie between intentions and performances, while initially plausible in the case of agentive abilities, will certainly not do the job when it comes abilities like smelling, digesting, or reading street signs without intending to read them. For abilities of that kind, the counterfactual will at best miss the point, and at worst come out straightforwardly false.

Taken together, these problems indicate very clearly, I should think, that the simple conditional analysis is on the wrong track. That is not to say that it may not serve as a starting point for a much more sophisticated view operating with counterfactual conditionals. But as it stands, it is definitely mistaken.

3 Possibilism

In the last chapter, we looked at the simple conditional analysis, which has its roots in the literature on voluntary actions and free will. In this chapter, we will look in detail at the other one of those two major views: the view that an agent has an ability to ϕ if and only if it is possible, in a properly restricted sense, for the agent to ϕ. I will refer to that view as "possibilism" in what follows.

Possibilism has its roots in the quest for a formal understanding of the modal auxiliaries, such as "must", "might", and – most crucially for our purposes – "can". The view emerged primarily in the course of the attempt to give a unified semantic and logical account of modal auxiliaries and has played a major role in the more formal regions of philosophy, as well as in logics and linguistics. Thanks to influential proponents, such as the linguist Kratzer (1977, 1981) and the logician von Wright (1951), possibilism is now the point of reference for most people with a background in those areas.[1] Interestingly, counterfactuals, or indeed conditionals in general, do not seem to play any role in those more formally oriented attempts to analyze ability statements. Instead, "can" statements in general, and *a fortiori* "can" statements expressing ability, are viewed as categorical statements about what is restrictedly possible for the agent to do.

To get a grip on possibilism, the natural place to start is Kratzer's modal semantics. In the next section, 3.1, I will give an outline of the crucial elements of that semantics, with a focus on "can" statements. In section 3.2, I will show how ability statements enter the picture; here, possibilism is introduced more formally. With the firm understanding of the view at hand, we can then move on to the discussion of possibilism. As we will see in sections 3.3 to 3.6, possibilism looks somewhat less problem-fraut than the simple conditional analysis in some ways, but, and this is the major import from the chapter, it ultimately fails as well – the truth conditions of ability statements are not properly stated in terms of restricted possibility. By the end of the chapter, we will have learned enough about abilities and the challenges in capturing them to be prepared to tackle the task of coming up with a superior view.

---

[1] Even though possibilism has its roots in the formal treatment of modal auxiliaries, it has not been limited to those areas. In the philosophical literature, proponents of the view have applied possibilism to the problem of time travel (Lewis 1976) and also, but less famously, to the problem of free will (Lehrer 1976; Horgan 1979), where possibilism has never quite managed to make it to center stage.

ᵭ OpenAccess. © 2020 Jaster, published by De Gruyter. [CC BY-NC-ND] This work is licensed under the Creative Commons Attribution-NonCommercial-NoDerivatives 4.0 License.
https://doi.org/10.1515/9783110650464-004

3.1 Modal semantics: Restricted possibility

In most areas of linguistics, "can" statements are taken to exhibit a unified semantics. According to that semantics, which has been developed in most detail by Kratzer (1977, 1981), "can" is a sentence operator which takes a proposition as its argument and forms a new proposition from it. "Pigs fly", for instance, can be taken as an argument for claims such as "Pigs can fly" or, in the negated form, "Pigs cannot fly". The structure of such statements is CAN(Pigs fly) and ¬CAN(Pigs fly), respectively.

What does the CAN operator do? It assigns possibility to the proposition over which it has scope. "Pigs can fly" states that it is possible that pigs fly. "Pigs cannot fly" states that it is not possible that they fly. CAN(Pigs fly) is thus equivalent to some statement of the form $\Diamond$(Pigs fly), where the diamond is the possibility operator.

What is it for something to be possible? To say that something is possible, according to the standard semantics, is for there to be a possible world in which that thing is the case. In other words, it is to apply existential quantification to the possible worlds. For it to be possible that pigs fly, say, there has to be a possible world in which pigs fly. To make a statement of the form CAN(p), then, is to state a possibility, the possibility that p. In possible worlds terminology, the truth conditions for such a statement are that there is a possible world in which p.

The kind of quantification that is applied to possible worlds is called the *modal force* of a modal expression. Expressions like "can" or "might" express possibility: the modal force of such expressions is existential quantification. For a possibility statement to be true, there has to be *at least one* world among the possible worlds in which the proposition over which the modal ranges is true.

Expressions like "must" or "ought", in contrast, express necessity. The modal force of such expressions is universal quantification. For a necessity statement to be true, *all* possible worlds have to be such that the sentence over which the modal ranges is true in them. Since we are concerned with Kratzer's semantics in connection with the idea that ability ascriptions express possibilities, the modal force we are going to be concerned with throughout the chapter will be existential quantification.

So much for the very basic framework. Let us now move on to the more complex bits. The complexity comes about in virtue of the fact that possibility statements (and modal statements in general) are virtually never unqualified – when we say that something is possible and thereby state that there is a possible world in which that thing is the case, we usually do not usually quantify over the totality of all worlds, but rather over a certain subset of those worlds – Kratzer calls this subset the *modal base* of the modal. In other words, possibility statements are usually statements of what we can call *restricted possibility*. When we think about whether p is possible, we are thinking about whether p is the case in some of the worlds that matter in our context. Let's call the worlds in the modal base the *relevant worlds*. (I'll use the terms "modal base" and "relevant worlds" interchangeably throughout the book.)

Which worlds are relevant is a matter of the conversational background, according to Kratzer. Where Kratzer speaks of the conversational background, I'll speak of a context. The idea is the same. It is simply that

> when a botanist says, in a foreign country with unfamiliar vegetation, 'Hydrangeas can grow here', she says not just that in some metaphysically possible worlds there are hydrangeas growing on this soil. She speaks only about worlds where the biology of hydrangeas, the geology and the climate of the country are as they are in actuality. When a detective says, 'Mary [cannot] be the murderer', she speaks about a different set of worlds: those that are compatible with everything she knows. In each case the conversational background selects a set of propositions; propositions about the actual facts of biology and geology, or the propositions that are known to the detective. The modal base is the set of worlds in which those propositions are true. (Vetter 2015: 68)

On Kratzer's view, then, "can" and other modals are actually used relationally: there is no absolute "can", but rather always a relational "can in view of". This relative modal phrase has two arguments: the first argument is provided by a phrase which specifies the facts in view of which it is true to say of someone that she can perform a given act – the set of propositions which determine the modal base. It is provided by the meaning of a phrase like "the biology of hydrangeas and the features of the soil", or "what the detective knows". The second argument is provided by the meaning of a sentence like "Hydrangea grow here", or "Fred is the murderer" (cf. Kratzer 1977: 341). It specifies what can be the case in view of the facts specified in the first argument.

The "in view of" locution will prove very handy in the course of the book. It is a good means to express, in ordinary language, that we are talking about a restricted realm of possible worlds. I will often use phrases of the form "S can φ in view of p", and when I do so, I use the "in view of" phrase in the very way Kratzer intends it: as a means to express that we are restricting the possible worlds in accordance to the facts specified by "p". The "in view of" locution provides the means to pick out the modal base worlds in a non-technical fashion.

Before we look more closely at some of the details of Kratzer's semantics, let's summarize the basic idea briefly. Kratzer offers an account of "can" statements, such as "Hydrangeas can grow here". Her fundamental insight is that any such statement carries a (typically implicit) "in view of p" supplement. To say that hydrangeas can grow here, in some context, is to say that they can, in view of some set of facts, grow on whatever place is referred to by "here" in that context. The "in view of" locution is to be interpreted in terms of *compossibility*. To say that hydrangeas can grow here is to say that the growth of hydrangeas on this spot is compossible with the features of the soil, the biology of hydrangeas, and whatever else matters to us in the context. To say that the two facts are compossible is to say that there is a relevant world – a world in which the soil and the biology of hydrangeas is as in actuality – in which hydrangeas grow.

A note on terminology. In what follows, I will often say that we *hold certain facts fixed* or *constant across possible worlds* or *modal space*. I find this a handy way of saying that certain facts go into the modal base, or, to put it differently still, that the relevant worlds are worlds in which those very facts obtain.

I will also often say of certain facts that they *get varied across the possible worlds* or *across modal space*. I use this locution as a means to put emphasis on the fact that certain features of the actual world are irrelevant for the determination of the relevant worlds, which form the modal base. When I say that hydrangeas can grow here, for instance, lots of features of the actual world are varied: my body weight, who the president of the United States is, and usually also whether or not there is a car standing right on the spot where it is said to be possible for the hydrangeas to grow. It is important to note that this talk about fixation and variation of facts does not add anything to Kratzer's framework, nor does it alter anything about it. It is merely a different way of putting things.

So much for the overview. Let us now proceed to some more subtle parts of the semantics. In particular, there are two aspect I would like to highlight, because they will prove helpful in embedding the possibilist semantics of ability statements in the possibilist semantics of "can" statements quite generally. The first is that there are generally two kinds of modal bases. Vetter's examples above illustrate this nicely. On the one hand, there are so-called *circumstantial* bases. Here, the relevant worlds are determined by propositions about facts in the world which do not depend on our epistemic access to them. The statement "Hydrangeas can grow here" is an example of a statement with a circumstantial base. The modal base contains facts about biology, the soil, and so on. "Pigs can't fly" can also be read as a statement about circumstantial possibility. It is naturally read as the statement that in view of their anatomy, their lack of wings and so forth, pigs cannot fly.

On the other hand, there are so-called *epistemic* bases. Here, the relevant worlds are fixed by facts we have knowledge about. "John cannot be the murderer" is naturally read along those lines. It is naturally read as the statement that, in view of what the detective knows about the crime and John's whereabouts, it is not possible that John be the murderer. Note that "Pigs can't fly" can also be read as a statement about epistemic possibility: in view of what we know about pigs, they cannot fly.

Both circumstantial possibility and epistemic possibility are *realistic*. That is to say that the actual world is always contained in the modal base. The reason for that is that the facts which determine which worlds count as relevant in a given context – the features of the world we hold fixed across possible worlds – are always that: features of the actual world. And since any world in which those facts obtain go into the modal base, the actual world will always automatically be among them. Epistemic possibility does not differ from circumstantial possibility in this respect. Since knowledge is factive – we can only know p, if p is true – the set of worlds in which the things we know are held fixed will always automatically contain the actual world.

Let's now move on to the second more subtle feature of Kratzer's framework. So far, I have introduced one important role of the context, or the conversational background, as Kratzer calls it: the context fixes a set of relevant worlds for possibility statements. Differently put, it determines a set of facts which are then held fixed across possible worlds. Something is possible, in the correspondingly restricted sense, if the so determined relevant worlds contain a world in which that thing is the case.

But this is not the only role the context plays in Kratzer's framework. The context also determines the so-called *ordering source*. As Kratzer observes, modal expressions are gradable. Some things are more easily possible than others. For some things there is a good possibility, for others merely a slight possibility. And so on.

The modal base is the wrong tool to account for this kind of graded modals. To account for the fact that there is a better possibility for Kim to be the murderer than there is a possibility for Jim to be the murderer, it will not do to check whether or not the modal base contains a world in which Kim is the murderer and whether or not it contains one in which Jim is the murderer. If each agent can be the murderer in view of what we know, say, then we will find both a world in which Kim is the murderer and one in which Jim is the murderer among the worlds in which we hold fixed what we know. To account for the fact that there is a better possibility for the one than for the other, we need a means to establish a scale of some kind.

The ordering source is designed to do just that (Kratzer 1981: 46 f.). It ranks the relevant worlds in accordance to their closeness to some ideal. Ordering sources come in different kinds. A bouletic ordering source ranks worlds by how close they are to fully satisfying the speaker's desires. A deontic sources ranks worlds by how close they come to meeting the demands of morality or the law. A stereotypical ordering source ranks worlds by how close they are to being entirely normal.

We can think of the ordering source as a set of propositions, which together express some ideal. Roughly, a world is closer to the ideal the more propositions of the ideal are true in it. A world in which all the laws are respected, for instance, is closer to the deontic ideal than a world in which some or all laws are broken.

That there is a better possibility for Kim being the murderer than there is one for Jim being the murderer will presumably be a matter of a *stereotypical* ordering source. The ideal is a world in which everything is normal. Worlds in which Jim is the murderer are less extraordinary than worlds in which Kim is the murderer. Thus, there is a better possibility for the former than for the latter, despite the fact that there is *some* possibility for each.

3.2 Possibilism: The details

Against the background of this overview, we can now return to our original topic: the understanding of ability statements in terms of restricted possibilities as it is suggested by possibilists. Probably the best-known expression of possibilism in the philosophical literature can be found in the following famous passage by Lewis:

> To say that something can happen means that its happening is compossible with certain facts. Which facts? That is determined (...) by context. An ape can't speak a human language – say, Finnish – but I can. Facts about the anatomy and operation of the ape's larynx and nervous system are not compossible with his speaking Finnish. The corresponding facts about my larynx and nervous system are compossible with my speaking Finnish. But don't take me along to Helsinki as your interpreter: I can't speak Finnish. My speaking Finnish is compossible with the facts considered so far, but not with further facts about my lack of training. What I can do, relative to one set of facts, I cannot do, relative to another, more inclusive, set. (Lewis 1976: 149)

In this passage, Lewis formulates the sketch of a view of abilities which treats ability statements as restricted possibility statements. What we do, when we ascribe the ability to φ to an agent, according to Lewis, is quantify existentially over a restricted domain of relevant possible worlds and postulate that the agent φ's in at least one of those worlds.

To say that ability statements express possibility is to say, in Kratzer's terminology, that the modal force of ability statements is that of existential quantification. This is the core claim of possibilism. The interesting question is: what is their modal base? What is obvious is that the modal base will have to be circumstantial, rather than epistemic. What matters for an agent's abilities is what the agent can do in view of certain facts in the world; the question is not what she can do in view of what we know about those or any other facts.

But which facts matter? Here, Lewis suggests a *contextualist* view: which facts go into the modal base – which features of the world are held fixed across modal space – depends on the speaker context. Sometimes, we may be interested in what Lewis can do in view of his anatomy. Sometimes, we may be interested in what he can do in view of his amount of training. Depending on the context, it will vary which facts go into the modal base.

Kratzer, whose work contains only a few scattered remarks specifically about ability ascriptions, suggests the very same picture. In the most explicit passage I know of, she considers the sentence "Ich kann nicht Posaune spielen" ("I cannot play the trombone") and writes:

> Depending on the situation in which I utter this sentence, I may say quite different things. I may mean that I don't know how to play the trombone. I am sure that there is something in a person's mind which becomes different when he or she starts learning how to play the trombone. A programme is filled in. And it is in view of this programme that it may be possible that I play the trombone. Or suppose that I suffer from asthma. I can hardly breathe. In view of my physical condition I am not able to play the trombone, although I know how to do it. I may express this by uttering ["Ich kann nicht Posaune spielen"]. Or else imagine that I am travelling by sea. The ship sinks and so does my trombone. I manage to get to a lonely island and sadly mumble ["Ich kann nicht Posaune spielen"]. I could play the trombone in view of my head and my lungs, but the trombone is out of reach. (Kratzer 1981: 54)

Kratzer and Lewis both endorse a contextualist account of ability statements, then. The modal base of ability statements varies in accordance to the context in which the statement is uttered. Again, the same picture is laid out by Horgan, an almost entirely forgotten proponent of possibilism. As Horgan writes in a brilliant paper on free will,

> [w]hen we say that (…) someone could have done something he did not actually do, we are saying that it was *possible* for the given (…) action to occur, even though it did not actually occur. I.e., we are asserting a statement of the form $\Diamond \phi$, where $\Diamond$ is the possibility-operator of modality. And according to contemporary semantical treatments of modality, $\Diamond \phi$ is true if and only if ϕ is true in some *possible world* that is *accessible* from the actual world. So if we wish to apply possible-world semantics to everyday "could"-statements, the key question we face is this: What is the relevant notion of accessibility? I suggest that the accessible possible worlds are those in which the circumstances are sufficiently similar to the circumstances which prevailed in the actual world. (…) The notion of circumstantial similarity is obviously quite vague (…). We resolve the vagueness (…) in varying ways, depending on the

context in which a given "could"-statement is used and the speaker's purposes in using it. (Horgan 1979: 346)

Lewis, Kratzer, and Horgan all voice the same idea: ability ascriptions have different modal bases, depending on the context of utterance.[2]

We are now in a position to formulate the possibilist's view a bit more precisely. In the beginning, we said that an agent has an ability, according to possibilism, if and only if it is possible, in a properly restricted sense, for the agent to ϕ. More specifically, though, possibilism is an explicitly *contextualist* view about ability statements: it is the view that an agent S has an ability to ϕ if and only if it is possible, in the properly restricted sense, for the agent to ϕ, *where the proper restrictions will vary across contexts*.[3] Or, to put it in possible worlds terminology,

> POSS. an agent S has an ability to ϕ if and only if there is a relevant world in which the agent ϕ's, *where the set of the relevant worlds will vary across ascriber contexts*.

It is the contextualist aspect of possibilism in particular, I take it, that makes the view extremely appealing. Ability statements do in fact seem to exhibit a certain context sensitivity that seems to have to do with our focus on certain features of the agent and facts of her situation. Can I jump the fence? Well, in a sense I can: I have done it many times in the past and it was easy. In a different sense, I cannot: my leg is currently broken. Can Emma kill her husband? In a sense, she certainly can: she owns a gun and knows very well how to use it. In a different sense, she cannot: she has tried it before when her husband was asleep and never managed to actually pull the trigger. Can Fred play the piano? In a sense, yes: he has had lessons since he was five years old. In another sense,

---

[2] One should not be distracted by the fact that Horgan operates with the notion of circumstantial similarity. As Lewis (1981) shows in connection with counterfactuals, there is no substantial difference between what he calls "ordering semantics" – a semantics which works with restrictions to the similar worlds – and "premise semantics" – a semantics which works with restrictions to worlds in which the relevant facts obtain.

[3] The explicitly contextualist formulation of the view adds something important to the original formulation: the original formulation leaves it open whether or not it varies which facts are held constant. Compare nomological possibility, for instance, to see this: p is nomologically possible if and only if there is a relevant world in which p is the case. But the relevant worlds will always be the worlds in which the laws of nature are held constant. In the case of abilities, there is no such general criterion to determine the relevant worlds. The overtly contextualist formulation of the view makes this fact explicit.

he cannot: he once had a blackout on stage and ever since his fingers grow stiff when he sits down to play.

These examples show very clearly that possibilism voices a very important insight. There is definitely something to the idea that abilities are always had *in view of certain facts*, to use Kratzer's terminology once more. It is in view of my stable intrinsic features that I can jump the fence, and in view of my broken leg that I cannot. It is in view of Emma's owning and mastering her gun that she can kill her husband, and in view of her personality that she cannot. It is in view of Fred's amount of training that he can play the piano, and in view of his trauma that he cannot.

Possibilism spells this out in terms of the restrictions we impose on the possible worlds. Sometimes, we hold only the agent's stable intrinsic features fixed and vary her current state; sometimes, however, her current state is held fixed as well. Sometimes, we hold only the agent's possession and her mastery of it fixed and vary her personality traits; sometimes, however, we hold the personality traits fixed, too. Sometimes, we hold only the agent's amount of training fixed and vary psychological impediments; sometimes, however, we hold the psychological impediments fixed as well. Depending on the modal base, the truth conditions of ability statements vary in the right way to account for our ordinary practice of ability ascription.

Possibilism is therefore not just attractive as a formal framework, which allows us to embed ability statements into a more general formal system encompassing "can" statements quite generally. Rather, it is also a way of understanding and systematizing the variances in the truth conditions of ability statements in ordinary language.

What about the ordering source? According to Kratzer, the ordering source is empty in the case of disposition statements, and the same, I take it, goes for ability statements (Kratzer 1981: 64). Whether this is really true need not concern us at the moment. We'll get back to this when we talk about degrees of abilities (→ 3.5). For now, let's just leave it at that.

We are now equipped with a rather solid understanding of possibilism and the formal framework underneath it. Thus armored, let us now dive into the discussion of the view. Here is a road map of the following sections.

The next two sections discuss a few up- and down-sides of possibilism. In the next section, 3.3, we'll see that the restrictions to varying modal bases will prove very fruitful in the handling of some of the issues that proved problematic for the simple conditional analysis. Thanks to the possibilist's fundamental insight that abilities are always had in view of certain facts and that the facts that matter vary across contexts, possibilism has no trouble at all with masks, for instance, and it provides a major leap towards an understanding of the difference between general and specific abilities. Here, the view has some obvious merits. In the course of the section, however, we will also see that possibilism clearly fails as an *account* of general abilities. The possibilist condition is simply insufficient for the having of such an ability. Just as the simple conditional analysis, possibilism exhibits a severe shortcoming in this regard.

Section 3.4 takes an equally balanced perspective. On the one hand, we are going to see that there are some obvious merits emerging from the fact that possibilism is somewhat simpler than the simple conditional analysis. Since possibilism does not focus on a modal tie between intentions and performances, but only on the possibility of the performance itself, possibilism does not have problems with impeded intentions and non-agentive abilities. Yet, we are also going to see that that simplicity comes at a price. The possibilist encounters serious trouble in delineating the boundary between abilities and other statements of restricted possibility. Possibilism may actually be a bit too simplistic when it comes to setting abilities apart.

Sections 3.5 and 3.6 focus on further problems with the view. In section 3.5, I will argue that possibilism fails to account for degrees of abilities and the context sensitivity of ability ascriptions that attaches to their gradable nature. In section 3.6, finally, I will side with Kenny (1976) in arguing that possibilism fails on formal grounds as well. Taken together, the problems outlined in the following sections are going to show very vividly that possibilism, while highly promising at the outset, turns out to be just as problematic as the simple conditional analysis on closer inspection.

3.3 Upsides & downsides I – General vs. specific abilities & masks

One reason why possibilism looks more promising than the simple conditional analysis at first is that the view seems to bring us a lot closer to understanding the difference between general and specific abilities and offers a neat understanding of the workings of masks. These are the upsides I will talk about in this section. The downside, however, will follow suit. As we will see in the last bit of the section, possibilism fails as an account of general abilities. The condition it states is far too weak here.

Let's start by focusing on the promising perspective on the distinction between general and specific abilities that is delivered by possibilism. According to the possibilist, an agent has an ability to φ if and only if there is a relevant world in which the agent φ's. The important notion with respect to the distinction between general and specific abilities is that of relevance: which worlds count as relevant in a given context varies. These variances, the possibilist can argue, generate abilities that are sometimes general, and sometimes specific.

Without using the terminology of general and specific abilities, Kratzer is quite explicit on the fact that she thinks of the distinction between general and specific abilities along those lines. Recall that, on her view, a speaker, call her Angelika, may say different things by uttering sentences like "Ich kann nicht Posaune spielen" ("I cannot play the trombone"). On the one hand, Angelika may express that she does not know how to play the trombone. On the other hand, she may express that the fact that she can hardly breathe or the fact that no trombone is available prevents her from doing it. As is easy to see, what Kratzer points at in distinguishing these uses of the sentence is the same distinction we recognized by the terms "general abilities" and "specific abilities". If one does not know how to play, one lacks a general ability to play. If one does not have a trombone available or has trouble breathing, one lacks a specific ability.

The way the distinction is accounted for is in terms of varying modal bases, or, put another way, of varying sets of relevant worlds. When we think about the general ability to play the trombone, we hold only certain intrinsic facts about Angelika's brain structure fixed – the "programme [that] is filled in" (Kratzer 1981: 54) when she learns how to play the trombone. When we think about the specific ability to play the trombone, in contrast, we also hold facts about her particular situation fixed: her breathing problems, for instance, or her current lack of a trombone.

This implies that, taken out of context, the question "Can Angelika play the trombone?" does not admit of a simple yes-or-no answer. In one sense, she can. In a different sense, she cannot. She has the general ability, but lacks the specific one. The possibilist does not seem to have trouble with any of this. Both Angelika's general ability to play the trombone and her lack of the specific ability to do it seem to be accounted for by the possibilist's framework very smoothly. They are both abilities in the sense that they can be accommodated by the same analysis. Quite generally, having an ability, according to the possibilist, is for it to be restrictedly possible for the agent to ɸ. General and specific abilities just differ with respect to the kinds of facts that go into their respective modal bases. Possibilism thus promises a unified account of both kinds of abilities.

By the same token, possibilism offers a neat understanding of masking. A masked ability, recall, is an ability which is present, but whose exercise would be interfered with. The interfering factor, we said, is a *mask*. My ability to swim may be masked by a cramp. My ability to sing a certain song may be masked by my hoarseness or my shyness or my forgetfulness of the lyrics. My ability to hit the bull's eye may be masked by my sudden dizziness.

The problem for the simple conditional analysis, recall, was that in masking cases, the counterfactual is false. It is not true that I would swim or sing or hit the bull's eye, if I intended to do so, because the closest intention worlds will be worlds in which the mask will be present. In the closest worlds in which I form the intention, I thus fail. Yet, I do have the relevant abilities. Hence, masking cases show that the truth of the counterfactual is not necessary.

Possibilism does not run into the same problem. When we are evaluating whether or not someone has the ability to swim, say, we are not moving into the closest intention worlds and seeing what happens there. Instead, we are restricting the possible worlds to the ones in which the relevant facts obtain and see whether or not there is one among them in which the agent swims. The important point with respect to masks is that, as long as the mask is not among the relevant facts, the modal base will comprise lots of worlds in which the mask is *not* present. Applied to our example: as long as we are not interested in the ability *to swim in view of the current cramp*, but simply in the general ability to swim – the ability to swim in view of one's fitness, say –, there will be quite a few worlds among the modal base in which the agent does not have a cramp and will thus swim. Possibilism is therefore quite apt to account for the fact that abilities can be had, and yet interfered with. They can be had in view of one set of facts, while their exercise would be interfered with by a different set of facts.

In chapter 2.2, when we talked about the problems the simple conditional analysis faces with respect to the distinction between general and specific abilities, we saw that the topic of general and specific abilities and the topic of masked abilities are connected. It is very obvious that general abilities can be masked, but unclear whether the same holds for specific abilities. Against the background of the preceding paragraphs, we can note that possibilism can make perfect sense of the controversy about the maskability of specific abilities. It promises a somewhat scalar account of more or less specific and general abilities. Let me explain.

Take the specific ability to do a handstand here and now. Does an agent who is, here and now, extremely scared of doing a handstand nevertheless have the specific ability to do a handstand here and now? Some may want to say she does – despite her fear, she has what it takes; her fear is just a mask. Others may want to say she doesn't. Given that she is as fearful as she is, she genuinely cannot do it. Her fear is not a mask, on that view, but actually deprives her of the specific ability.

For a similar example, let us ask: does the agent have or lack the specific ability to do a handstand here and now if, although nothing inhibits the agent's exercise of her general ability to do a handstand on the macro level, the micro facts happen to be such that something will interfere with the agent's exercise of doing a handstand? Suppose, for instance, a hyperacidity prevails in the agent's muscles so that any attempt to do a handstand is bound to fail due to a cramp in the agent's biceps. Does this deprive the agent of the specific ability to do a handstand here and now? Or does it just mask it? I take it that people will disagree about such questions.

For the possibilist, this should not come as a surprise. What it suggests is that there is actually no such thing as *the* correct understanding of specific abilities. The disagreements we have just considered can be solved only by *re*solving them. If we count only worlds as relevant in which absolutely everything, including the hyperacidity in the agent's muscles, is as in actuality, then it will be correct to say that the agent lacks the ability to do a handstand. If the hyperacidity is not among the facts we hold constant, we will correctly ascribe the ability to her; the hyperacidity will be a mere mask. Likewise in the first example. If the agent's fear is not among the relevant facts, we will ascribe the ability and count the fear as a mask. If it is, the fear deprives the agent of the ability.

Whether or not a specific ability is maskable will therefore depend on the comprehensiveness of the modal base. If absolutely every fact of the agent's situation is held fixed across modal space, then any impediment to the exercise of whatever act we are considering will itself have to be held fixed. Hence, the agent's ability to do a handstand is not *masked* by the agent's fear or a hyperacidity in her muscles, but *annihilated*. On this interpretation of a specific ability, then, specific abilities do not allow for masks. The modal base narrows down to exactly one world: the actual one.

If, in contrast, we are thinking of specific abilities in such a way that some facts of the agent's situation may still vary, then there is room for masks. For now we can abstract away from the fear and hyperacidity, say, and vary them across the possible worlds. And once we allow for that, the agent may very well come out as having the specific ability to do a handstand, despite her actual fear or her actual hyperacidity.

The picture this suggests is that abilities can be more or less specific, depending on just how comprehensive the set of facts is that determines the relevant worlds among which we then look out for a world in which the agent performs whatever act it is we are interested in. There are maximally specific abilities on one pole – for those abilities, we have to hold fixed absolutely every feature of the situation, be it ever so tiny. Along the scale, we can then align less specific abilities – abilities which allow for more and more variance of features of the agent's particular situation. The disagreement about the maskability of specific abilities can be resolved that way. Once we see that the two opposing parties in the disagreement focus on varyingly extreme kinds of specific abilities, their disagreement vanishes into thin air.

This exposition has been quite simplistic, and much more would have to be said about a scalar understanding of general and specific abilities. Since I myself endorse a treatment of the distinction between general and specific abilities along very similar lines, we will have lots of time to turn to that task later on (→ 4.5). For now, let's just note that the possibilist has a very interesting story to tell about the relation between general and specific abilities. In virtue of its appeal to varyingly comprehensive modal bases, it accommodates both and allows for a scalar picture of with maximally specific abilities on one pole, less than maximal, but yet specific abilities in the middle and general abilities towards the other end of the scale.

So much for the upside of possibilism with respect to the distinction between general and specific abilities and the understanding of masks. Now on to the downside. Although possibilism seems to capture one important aspect of the difference between general and specific abilities, we will now see that possibilism fails as an *account* of general abilities. More specifically, the problem is that the possibilist condition is not sufficient for the having of a general ability. It is much too easily met.

One way to approach the problem is to note that the modal base of ability statements is always realistic, according to the possibilist. Since what we hold fixed across modal space are always features of the actual world, the actual world is necessarily part of the modal base; it is always among the set of relevant worlds. To illustrate, suppose we are interested in an archer's ability to hit the bull's eye. If we are interested in her general ability, we hold certain features of the archer herself fixed: her brain structure, her vision, her muscular constitution maybe. What we do not hold fixed is her broken arm, say. If we are interested in the specific ability, we hold the broken arm fixed, too. Regardless of whether we are interested in the general or the specific ability, then, the relevant worlds will be those in which certain facts – states of affairs that obtain in the actual world – obtain. And of course, the actual world will be among those worlds. It will be a member of the set of worlds in which a certain subset of the actual facts obtain.

Obviously, this yields a problem. If the actual world is always among the relevant worlds, then any agent will automatically have the ability to do whatever she does in the actual world. To see that, just recall that to have an ability to φ, according to the possibilist, is for there to be a relevant world in which the agent φ's. And given that the actual world is a relevant world, this condition is obviously met if the agent φ's in the actual world.

The problem with this is that a singular successful performance of some act ϕ, be it in the actual world or in any of the merely possible worlds, will rarely, if ever, suffice to show that an agent has a general ability to ϕ. If I hit the bull's eye in the actual world, this does not show anything at all about my general ability to hit the bull's eye. This shows that I had the (maximally) specific ability to hit the bull's eye, certainly. But clearly, a singular success does not show that the agent has a general ability. This is a serious problem for the possibilist. The view apparently misses out on an important dimension here.

It is important to see that even though I have approached the problem from the assumption that the modal base always comprises the actual world, the problem itself has nothing to do with that particular feature of possibilism. Rather, it has to do with the existential quantification over the modal base quite generally. That is why I said in the beginning that the problem with possibilism is the modal force it assigns to ability statements. It is the single most essential feature of possibilism – the commitment to an analysis of abilities in terms of possibility – that causes the trouble.

To see that this is so, note that just as a singular successful performance of ϕ in the actual world suffices to make it true that the agent ϕ's in a relevant world, a singular successful performance of ϕ in any other single one of the worlds that comprise the modal base is equally insufficient for an agent to have a general ability to ϕ. Yet, possibilism predicts otherwise. Once there is a successful performance of ϕ to be found in the modal base worlds, the agent comes out as having the ability. Apparently, this is not a satisfactory account of general abilities.

The upshot of this section is mixed. On the one hand, possibilism clearly has a few merits over the simple conditional analysis in that it, first, offers an understanding of one important dimension on which general and specific abilities seem to differ and, secondly, yields an understanding of the workings of masks. On the other hand, possibilism does not really seem to capture what it is for an agent to have a general ability. The condition stated by the possibilist is not sufficient for an agent having a general ability.

3.4 Upsides and downsides II – Impeded intentions, agentive vs. non-agentive abilities, and other possibilities

In the last section, we saw that possibilism has one very important advantage over the simple conditional analysis. In this section, we will turn to a second feature that is both a merit and a problem for the view. Possibilism is very simple. It analyzes the ability to ϕ simply in terms of the possibility of ϕ-ing and not, as the simple conditional analysis, in terms of a modal tie between the agent's intentions to ϕ and her performances of ϕ-ing. Consequently, possibilism does not run into problems having to do with the postulation of such a tie. Specifically, it does not have the problems with impeded intentions or non-agentive abilities. This should not be overvalued, however. For as I will argue, possibilism may in fact be *overly* simplistic in that it fails to explain what it is that unifies abilities and sets them apart from mere physical possibilities, say.

That possibilism does not run into the problem of the impeded intention is easy to see. In cases of an impeded intention, recall, the agent is unable to ϕ in virtue of an impediment that prevents her from intending to ϕ in the first place. One of the examples I used to illustrate the problem was that of the coma patient (van Inwagen 1983: 119) who cannot raise her arm in virtue of being in a coma. The coma does not just prevent the agent from raising her arm, though, it also prevents her from intending to raise her arm to begin with. And this caused trouble for the simple conditional analysis.

Since the simple conditional analysis analyzes abilities in terms of a counterfactual tie between intentions and performances, to find out whether or not the agent has an ability, we had to move into the closest intention worlds and see whether or not the agent raises her arm there. But since the closest intention worlds are not, in fact, coma worlds, the counterfactual comes out true. If the agent intended to raise her arm, she would, because she would not be in a coma in that case. That is the wrong result.

The possibilist does not have the same problem. The possibilist restricts the possible worlds to the ones in which the relevant facts obtain. And among those, she checks for performance worlds. In virtue of this feature of the view, cases of impeded intentions are unproblematic. In the case of the coma patient, the coma will usually be among the relevant facts. Usually, we want to know what the coma patient can do *in view of her coma*. Is there an arm raising world among the so restricted worlds? Of course not. The modal base will neither contain arm raising worlds, nor will it contain any worlds in which the agent intends to raise her arm to begin with. The worlds the possibilist looks at all contain the very impediment that prevents both the intention to ϕ as well as ϕ-ing. Thus, there will not be ϕ-ing worlds among the relevant worlds. And hence, possibilism rightly yields that the coma patient and other agents whose intentions are impeded lack the ability to raise their arms, or whatever it is they cannot do in virtue of that impediment. Clearly, possibilism has an advantage over the simple conditional analysis here.

What about the good sense in which the coma patient *can* raise her arm, though? Recall that when we talked about impeded intentions in connection with the simple conditional analysis (→ 2.3), we said that there is a good sense in which the coma patient retains quite a lot of her abilities – despite her coma. The ability to raise one's arm is not the best example to see that. But when you think of abilities that were acquired by a lot of training, like juggling, for instance, you will come to sense that there is a rather good sense in which one will want to say that the coma patient has retained that ability.

Possibilism has no trouble with any of this. In fact, it can account for both the sense in which one will want to say that the coma patient can, and the sense in which one will want to say that she cannot raise her arm, juggle, or whatever one is interested in. The key feature of possibilism here is the variance in the modal base. Sometimes, we are interested in what the agent can do in view of her coma. In that sense, she cannot do much. Sometimes, we are interested in what she can do in view of her amount of training in a certain discipline, or her muscular constitution, or a certain "programme in her brain", as Kratzer (1981: 54) has put it. In that sense, the coma patient has the ability to do quite a lot of things.

The possibilist can explain that smoothly. When we are interested in what the coma patient can do in view of her coma, the modal base will comprise only coma worlds. And among those, there is no arm raising world and no juggling world. But when we are interested in what the coma patient can do in view of her amount of training in something, the coma is varied across the possible worlds and the modal base will comprise all kinds of worlds in which the agent is not, in fact, in a coma. Among the so selected set of worlds, there will be quite a few in which the agent raises her arm and juggles. The possibilist is therefore apt to account for both senses of "having an ability" here – the sense in which the coma patient can, and the sense in which she cannot do certain things.

One last note on this, just to tie things back to the topics we have touched upon earlier. All this can also be put in terms of masks. Depending on how we select the modal base, the coma will either be a mask of the agent's ability to juggle, say, or it will deprive her of it. When the modal base comprises only coma worlds, the coma is an annihilator of the ability – the agent cannot juggle. When the coma is varied, the coma is a mask – the agent retains the ability to juggle, despite being in circumstances that conflict with the exercise of that ability. Things fall into place quite neatly here. Cases of impeded intentions are smoothly accommodated by possibilism.

Now on to non-agentive abilities. Here, too, possibilism looks more promising than the simple conditional analysis. Again, this has to do with the fact that intentions do not play a prominent role in the possibilist account. Non-agentive abilities, recall, include the ability to see, smell, or hear; to digest, dissociate, or produce saliva; to unintentionally read street signs or to understand something without listening. All of those things can, or can only, be exercised without a foregoing intention on the agent's part. This spelled trouble for the simple conditional analysis (→ 2.6).

Possibilism does not have the same problem. According to the possibilist, an agent has an ability if and only if there is a relevant world in which the agent φ's. This condition works equally well for non-agentive as for agentive abilities. An agent has the ability to bake a cake if and only if there is a relevant world in which she bakes a cake. An agent has the ability to produce saliva if and only if there is a relevant world in which she produces saliva. An agent has the ability to unintentionally read street signs if and only if there is a relevant world in which she reads street signs unintentionally. The fact that saliva is usually produced without a foregoing intention and street signs *can* only be read unintentionally, if no foregoing intention is formed, does not cause a problem at all for the possibilist. The modal base can comprise intention worlds and non-intention worlds alike. Intention worlds do not enjoy any special status.

So much, again, for some of the upside of possibilism. Let's now turn to a downside once more. The downside is that the simplicity of possibilism comes at a price. It is hard to see what unifies abilities and sets them apart from other restricted possibilities, such as mere physical possibility, say, on the possibilist framework. There is good reason to suspect that possibilism may in fact be a bit *too* simplistic; there seems to be an important dimension of what it is for an agent to have an ability that eludes the view.

Nomological possibility is possibility in view of the laws of nature, epistemic possibility is possibility in view of what is known, deontic possibility is possibility in view of what is allowed. Apart from those sharply contoured possibilities, there are other possibilities, of course. Possibility in view of certain facts of a situation, for instance. It may be possible in view of the security measures that a rollercoaster car derails, but not in view of the actual condition of the car and the rails. It may be possible in view of the closeness of a cliff that you fall off the mountain you are standing on, but not in view of your degree of carefulness.

If possibilists are right, then the possibility involved in the latter two possibility statements resembles that of ability ascriptions very closely. In both cases, we have existential quantification over a circumstantial modal base. Yet, "the car can derail" and "you can fall of the cliff" are not ability ascriptions, but rather ascriptions of what is often called a "physical possibility". A natural question to ask the possibilist is: what is the difference?

I don't think the possibilist has a fully convincing response to this question at hand. I guess what possibilists *should* reply is that the kind of *possibility* involved in ability ascriptions and statements like "the car can derail" and "you can fall off that cliff" is actually the same. That is the essence of their view about abilities, after all. They could then go on to try and characterize ability ascriptions in terms of other features of such ascriptions. One way one may try to do so is by characterizing ability ascriptions as statements that express restricted circumstantial possibility statements which relate agents to actions. "The car can derail" and "you can fall off that cliff" is not an ability ascription, according to this characterization, whereas "the president can start a war" is.

In view of the fact that there are such things as non-agentive abilities, however, it should be obvious that this characterization won't do. For possibilism to cover non-agentive abilities as well, the possibilist can hardly characterize ability statements by postulating that they relate agents to actions. Smelling, digesting, and reading street signs unintentionally are not (typically) actions. Yet, one can have the ability to do these things. Abilities will therefore have to be individuated differently, and it is hard to see how the possibilist is supposed to tackle this task with the tools available to her. Perhaps it can be done. But there is definitely an open question here.

3.5 The problem with degrees and context sensitivity

So far, we have noted both merits and problems of possibilism. In this and the next section, it will become obvious that the debits problems preponderate. In this section, we will see that possibilism fails to account for degrees and the corresponding kind of context sensitivity that attaches to ability statements. In the next section, we will see that the view is formally inadequate in that the most basic laws governing possibility fail to hold for ability statements.

Abilities, we said (→ 1.3), come in degrees. Glenn Gould has the ability to play the piano to a higher degree than I do; my grandmother sews better, my cousin drives better, my father draws better than I do. One meteorologist may be better at predicting the weather than another. Two archers may be varyingly good at hitting the target.

Looking more closely at degrees of abilities, we saw that the degree of an agent's ability can be influenced on two dimensions: the degree of an ability depends, first, on the quality of the agent's performances and, secondly, on the range of circumstances across which the agent manages to deliver those performances. We called those two dimensions *the dimension of achievement* and *the dimension of reliability* respectively.

We also said that the gradable nature of abilities results in a specific kind of context sensitivity of ability ascriptions. That is because the degree of an agent's ability to φ required for the agent to count as having the ability to φ *simpliciter* varies across contexts. A darts player who misses 99% of the time, but manages to hit the bull's eye every 100th time does not usually count as having the ability to hit it. There may be contexts in which we count her as having the ability, though. She has what it takes; it is within the range of what she can do. The context, we said, determines some threshold along the scale of degrees of abilities above which an agent counts as having the ability *simpliciter*. Once we have degrees, then, we also have the mechanism for the corresponding context sensitivity.

The problem with possibilism is that it fails to account for degrees and, consequently, for the corresponding kind of context sensitivity of ability ascriptions. As in the case of the simple conditional analysis, the problematic cases are not so much the ones in which achievement is the dimension which matters, even though they, too, require a supplementation of the possibilist framework as it stands. To see that, reconsider the driving abilities of my partner and his friend. Both of them can drive a car, but while my partner is a very good driver, his friend drives poorly. At first sight, this is a problem for the possibilist. The possibilist holds that an agent has an ability to φ if and only if there is a relevant world in which the agent φ's. And obviously, this is true for my partner and his friend alike. His friend can drive, after all. It is thus possible, given her amount of training and so forth, that she drives. The problem is that she is not very good at it. So, her driving performances will be of rather low quality. As it stands, possibilism does not have the means to account for this kind of difference between two agents.

Of course, the possibilist framework can easily be supplemented by an account of degrees that track variances on the level of achievement. The possibilist can avail herself of a strategy analogous to the one I suggested in connection with the simple conditional analysis (→ 2.5). She can attach values to the quality of an agent's performance and postulate that the degree of an agent's ability is higher the higher that value is. Were achievement the only dimension that counted, an account of degrees of abilities would be quite easily available for the possibilist.

But achievement is not the only dimension that counts. To come up with a comprehensive account of degrees of abilities, variances on the dimension of reliability will have to be accounted for as well. This is where possibilism exhibits a deep structural shortcoming. Reconsider Ted and Fred, our two meteorologists. Fred has developed an algorithm, by means of which he can predict 99% of all weather constellations correctly. Ted has developed a faulty algorithm – it goes wrong for 45% of all weather constellations. Clearly, Fred is the better meteorologist. Possibilism, however, has no means to distinguish between the degrees of Ted's and Fred's weather prediction abilities. For both, there will be relevant worlds in which they predict the weather correctly. According to the possibilist, the two are on a par.

The problem with possibilism is that the existential quantifier does not seem to be particularly well-suited to account for degrees. Once there is *one* performance world among the relevant worlds, the possibilist condition is satisfied. There does not seem to be any room for the kind of scale that would have to be established in order to account for differences on the dimension of reliability.

But wait a minute! Did we not talk about degrees of possibility earlier on? And wasn't there a tool that Kratzer employed in her semantics to account for such degrees? – The tool that comes to mind here, of course, is the ordering source I mentioned when I first introduced the standard semantics of "can" statements (→ 3.1). And indeed, it seems to be just the right means for the purpose at hand. It is designed to account for degrees of possibility. Why not use it in order to account for degrees of ability, too?

Unfortunately, the ordering source does not do the job. The ordering source, recall, ranks the possible worlds according to some ideal. One very common ordering source is a stereotypical one. In the stereotypical ideal, everything is normal and the possible worlds are then ranked in accordance to their similarity to that ideal – the more normal a world is, the closer it comes to the stereotypical ordering source. We saw that a stereotypical ordering source is what is at work when it comes to the detective's judgement that there is a better possibility of Kim being the murderer than of Jim being the murderer. Each of them can be the murderer, but the closest worlds in which Kim is the murderer are much closer to the ideal of normality than the closest worlds in which Jim is the murderer.

So far, so good. The crucial question is: can the same kind of strategy be employed for degrees of ability? Is there an ordering source, such that the closest worlds in which Fred predicts the weather correctly are always closer to the ideal than the closest worlds in which Ted predicts the weather correctly? At first, it may seem plausible that the stereotypical ordering source is already the right one for the purpose at hand. For are not worlds in which Fred predicts the weather correctly more normal than worlds in which Ted predicts the weather correctly? After all, Fred is far more reliable in his weather predictions than Ted.

But we need to be careful. What we need to ask is: what would be the case in the most normal world imaginable? Then we have to find out how close to that world the closest world is in which Fred predicts the weather as opposed to the closest world in which Ted predicts the weather. The answer suggests that the stereotypical ordering source will not do the job of telling Fred's and Ted's respective degrees of the ability to predict the weather apart. For it seems that in the most normal world imaginable, both Fred *and* Ted will predict the weather correctly. Both have a quota of correct predictions that exceeds 50%, after all. Thus, it would seem that a world in which both agents predict the weather correctly is, *ceteris paribus*, more normal than any world in which one of them makes a wrong prediction.[4] Hence, the closest worlds in which Fred makes a right prediction is just as close to the ideal as the closest world in which Ted makes a right prediction.

From here, we can move in two directions. We can either try and come up with all kinds of ordering sources and see whether one of them is suited to account for degrees of agents' abilities. Or we can think about the prospects of solving the problem of degrees by the use of ordering sources quite generally. I suggest we do the latter. The ordering source as such fails to provide a proper means to account for degrees of ability.

To see that, note that reliability is a matter of performances across various situations – it is a matter of a quota. If I am more reliable than you are in hitting a target, then this has to do with my hitting the target across a wider range of circumstances. The possibilist, however, can only ever take one world into account. Even if we rank the worlds in accordance to some ideal, then, – stereotypical or of any other kind – we will still have to judge the agent's degree of ability on the basis of her performance of just one world, namely that which comes closest to that ideal. Given the nature of reliability – its tight connection to the idea of a quota – this strategy seems faulty in a very substantial way. No matter what kind of ranking we impose on the possible worlds, the fact that there is a difference between two agents in the worlds closest to the ideal will never ensure that this difference mirrors the agents' differences in terms of some relevant quota across worlds.[5]

3.6 A formal problem

Up to this point, several doubts concerning the viability of a possibilist account of abilities have accumulated. Possibilism is unable to set abilities apart from mere physical possibilities (→ 3.4), it fails to account for general abilities (→ 3.3), and it sits more than awkwardly with the idea of degrees of ability (→ 3.5). In this section, we will confirm the impression that there is indeed something wrong with the view. As I will lay out in what follows, possibilism is actually formally inadequate. There are formal considerations which show rather clearly that the "can" of ability is not the "can" of possibility.

---

[4] We can stipulate that the predictions of both agents are better the more normal the weather conditions are.
[5] Vetter (2015) argues against a possibilist understanding of degrees of dispositions along the same lines.

According to the possibilist, the "can" of ability functions as a restricted possibility operator. But as Kenny (1976) has convincingly shown, the "can" of ability cannot be accounted for in terms of a possibility operator, because the "can" of ability does not obey the most fundamental laws of the logic of possibility. Since the possibility operator is defined by the formal rules that govern it, this is a knock-down argument. It just does not make any sense to claim that the "can" of ability is a restricted possibility operator, if none of the most basic principles of the possibility operator apply. Let us go through Kenny's argument to see how it works.

Suppose the possibilist is right in her verdict that ability statements, such as "S can ϕ", express something along the lines of "It is possible (in some properly restricted sense) for S to ϕ". Then the logical form of ascriptions of ability ascriptions would be CAN(S ϕ's), where CAN is a properly restricted possibility operator. If CAN is a possibility operator, then it will have to be construed as a sentence operator. And the sentence it ranges over will express a proposition according to which some subject engages in some doing, paradigmatically an action. All of that should be familiar from the section on the basic Kratzer semantics (→ 3.1). On this construal, ability ascriptions would be akin to other forms of restricted possibility statements, such as deontic, nomological, or epistemic possibility claims.

The problem Kenny notes is that ability statements do not exhibit the right kind of logic to pass for possibility statements; they fail to obey the most basic laws of possibility. To see that, note that the weakest system of modal logic suitable for alethic interpretations of the operators is the system KT. One axiomatization of KT contains (among other principles) the axiom schemata

T. p → ◊(p)

K. ◊(p) ∨ ◊(q) ↔ ◊(p∨q)

The validity of these two principles can be derived directly from the possible worlds semantics of the possibility operator. In possible worlds terminology, the T-axiom reads: if p is the case in the actual world, then there is some world in which p is the case. That is clearly valid, because the actual world, of course, is some world.

The K-axiom can be split up into two conditional claims. From left to right it states that

K1. ◊(p) ∨ ◊(q) → ◊(p∨q)

From right to left it states that

K2. ◊(p∨q) → ◊(p) ∨ ◊(q)

In possible worlds terminology, K1 reads: if there is some world in which p is the case or there is some world in which q is the case, then there is some world in which p or q is the case. Apparently, this is a logically valid principle, for "p or q" will hold both in any p-world and in any q-world. Hence, if there is some p-world or some q-world, *eo ipso*, there will be some p-or-q-world.

Likewise for K2. In possible worlds terminology, the principle reads: if there is some world in which p or q is the case, then there is a world in which p is the case or there is a world in which q is the case. Again, this principle is clearly valid. Since any p-or-q world is either a p-world or a q-world, there being a p-or-q-world implies that there is a p-world or a q-world.

As Kenny points out, neither of these schemata is correct, if the possibility operator (the diamond) is interpreted as the "can" of ability. His arguments proceed by counterexample. Let's start with the T-axiom p → ◊p. Interpreting the possibility operator in terms of the "can" of ability, the theorem states that

T_CAN. If S ϕ's, then S can bring it about that S ϕ's.

Note that I am using the formulation "S can bring it about that S ϕ's". This locution has the purpose of reflecting, in ordinary language, that the "can" of ability functions as a sentence operator on the possibilist's framework. Nothing substantial hinges on this. It just helps to integrate natural language ability talk into the schemes of the modal laws governing possibility.

On the face of it, T_CAN looks like a valid principle. As Geach points out, it seems to be a truism that "what a man does he can do; that is clear if anything in modal logic is" (1957: 15). And according to Mele,

> there is an ordinary sense of 'able' according to which agents are able to do whatever they do. In this sense of 'able', an agent's having A-ed at a time is conceptually sufficient for her having been able to A. (Mele 2002: 447f.)

Suppose Geach and Mele are right. Suppose, that is, that there is, in fact, an ordinary sense of the "can" of ability which is in line with axiom T.[6] The problem is just that there are other senses of the "can" of ability which are clearly not. For

---

[6] You may struggle with the idea that there is such a sense. I do not want to commit on this issue. As it will turn out, my own view has the resources to account for a sense of "specific ability", according to which a subject has the specific ability to do whatever she did. But my view does not commit me to endorsing this sense (→ 4.5, b & d).

general abilities, for instance, the principle clearly fails. As Kenny points out, for instance,

> "[a] hopeless darts player may, once in a lifetime, hit the bull, but be unable to repeat the performance because he does not have the [general] ability to hit the bull. (...) Counterexamples similar to [this] will always be imaginable whenever it is possible to do something by luck rather than by skill. But the distinction between luck and skill is not a marginal matter in this context: it is precisely what we are interested in when our concern is [general] ability, as opposed to logical possibility or opportunity" (Kenny 1976: 214).

What Kenny points out here is that the having of a general ability requires more than just an individual success. This insight is not new; I emphasized the same fact in the section about general and specific abilities (→ 3.3), when I argued that possibilism fails to account for general abilities. The new twist to this insight is that possibilism is therefore at odds with the logic of possibility. The T-theorem, while valid for all established alethic interpretations of the possibility operator, is clearly invalid for the "can" of general ability. This fact casts additional doubt upon any view according to which the "can" of general ability is understood in terms of the possibility operator.

Let's now turn to the K-axiom, which seems likewise problematic, if the diamond is interpreted as the "can" of ability. As Kenny points out, it fails in both directions, but in what follows, I will focus on K2.[7] Inserting the "can" of ability for the diamond, K2 reads

> K2_CAN. If S can bring it about that S ϕ's or ψ's, then S can bring it about that S ϕ's or S can bring it about that S ψ's.

On the face of it, this, too, looks like a valid principle, and indeed, it has many true instances. If I can bring it about that I ride my bike or drive my car, then I can bring it about that I ride my bike or I can bring it about that I drive my car. If I can bring it about that I eat an apple or a banana, then I can bring it about that I eat an apple or I can bring it about that I eat a banana.

---

[7] Interpreting the diamond as the "can" of ability, K1 states that if S can bring it about that S φ's and S can bring it about that S ψ's then S can bring it about that S either φ's or ψ's. This is false, as can be seen when ψ is non-φ. The president has the ability to bring it about that the president destroys Moscow and he has the ability to bring it about that the president does not destroy Moscow. Yet, he doesn't have the ability to bring it about that the president destroys Moscow or the president does not destroy Moscow. Why does he lack that ability? Because (The president destroys Moscow or the president does not destroy Moscow) is a necessity, and according to Kenny no one has the ability to bring about a necessity.

But there are problem cases. Suppose I am faced with a stack of cards that is turned face-down (Kenny 1976: 215). In that case, I have the general ability to pick red or black. Yet, it seems, I neither have the general ability to pick red nor do I have the general ability to pick black. It is therefore false that

> #. If S can bring it about that S picks black or red, then S can bring it about that S picks black or S can bring it about that S picks red.

For an even better example, suppose I have the general ability to hit the dart board (ibid: 215f.). This implies that I have the general ability to hit the upper or the lower half of the dart board.[8] Yet, I may well lack the general ability to hit the upper half of the bull and also lack the general ability to hit the lower half. It is therefore false that

> ##. If S can bring it about that S hits the upper half or S hits the lower half, then S can bring it about that S hits the upper half or S can bring it about that S hits the lower half.

Why is this an even better counterexample to K2_CAN? Because it can be strengthened. The example works better the more discriminatory we get. That I can hit the dartboard implies that I can hit one square-millimeter on the board, or another, or another, and so on. But it obviously does not imply that I can hit any particular square-millimeter on the board.

The important point is that in both examples, the left-hand side of the conditional is satisfied while the right-hand side is not. And, as Kenny points out, "[s]imilar counterexamples can be constructed in connection with any other discriminatory skill" (ibid. 215). Kenny has a very strong point here. The "can" of ability disobeys the rules of the possibility operator. This makes it almost impossible to cling to the view that the "can" of ability is a "can" of possibility.

3.7 Upshot

Let's go through the main line of thought of this chapter once more. In this chapter, we looked at possibilism – the view that an agent has an ability to φ if and only if it is restrictedly possible for the agent to φ; or, to put it in the possible

---

[8] If you don't like the inference from "having the ability to hit the board" to "having the ability to hit the upper half or having the ability to hit the lower half", never mind. We can run the argument by assuming right away that the agent has the latter ability.

worlds terminology we used throughout the chapter, if and only if there is a relevant world in which the agent ϕ's. We noted quite a few merits about possibilism, but also a number of problems, which together showed very clearly that abilities are not, in fact, restricted possibilities.

Let's start with the merits. First of all, possibilism seemed much better suited than the simple conditional analysis in making sense of the distinction between general and specific abilities (→ 3.3). The crucial tool here was the variance in the set of worlds that go into the modal base: having a general ability, on the possibilist framework, is for there to be a ϕ-ing-world among *one* set of worlds, whereas having a specific ability is for there to be a ϕ-ing-world among a *different* set of worlds. When I have a broken leg, say, I still have the ability to dance in view of the "dancing program" in my brain, but not in view of my broken leg. And depending on whether or not we hold the broken leg or merely the dancing program fixed, there either will or won't be a world among the thus-restricted set of worlds in which I dance. This looks very promising at the outset and seems to capture a very important dimension on which general and specific abilities differ.

By the same token, possibilism allows for an understanding of the workings of masks (→ 3.3). The broken leg is a mask for the general ability to dance. It is a fact in the world that would interfere with the exercise of the ability. Yet, it does not deprive the agent of the ability, because it is not among the set of facts that are held fixed across the possible worlds. When we are interested in the general ability to dance, we are not interested in the ability to dance in view of one's currently broken leg, but rather in the ability to dance in view of the dancing program in one's brain or one's overall bodily constitution, or what have you. Yet, the broken leg clearly deprives us of the more specific ability to dance in view of one's broken leg. Here, the broken leg is held fixed, and hence there is no dancing world among the modal base worlds. Possibilism explains all this very smoothly.

The third merit we noted was that possibilism does not run into the problem of the impeded intention (→ 3.4). The coma patient cannot raise her arm, on the possibilist framework, because the coma will have to be held fixed across the possible worlds and hence there will not be any arm raising worlds among the modal base. Since the possibilist does not put emphasis on the link between intentions and performances, there is no need to move beyond the coma worlds in order to reach the closest intention worlds. Instead, the possibilist checks for performance worlds among the coma worlds. And voilà: there are none. This is the right result. Again, that is not to say that there is no sense in which the coma patient retains certain abilities. The possibilist can account for both senses, depending on whether or not the coma is held fixed.

Fourth, and relatedly, possibilism does not run into problems with non-agentive abilities (→ 3.4). Since abilities, quite generally, are understood in terms of the possibility of performance, non-agentive abilities do not pose any particular problem. Having the ability to dance is for it to be possible for the agent to dance. Having the ability to produce saliva is for it to be possible for the agent to produce saliva. Since no emphasis is put on the tie between intentions and performances, no problem emerges from abilities that can or can only be exercised without a foregoing intention on the agent's part.

But there were also some severe problems with the view.

First, there was a problem with general abilities (→ 3.3). Possibilism may shed light on one dimension on which general and specific abilities differ, but it certainly fails as an account of general abilities. Having a general ability to φ seems to require a certain quota of success in φ-ing. There being just one world (actual or possible) in which the agent φ-s certainly isn't enough. The last word on general and specific abilities obviously has not been spoken by the possibilist.

Secondly, the possibilist fails to properly delineate abilities from other possibilities, like physical possibilities (→ 3.4). What is the difference between a statement of the form "I can dance" or "I can read street signs unintentionally" and a statement of the form "You can fall off the cliff"? All three express restricted circumstantial possibility. You and I are both agents. And reading street signs unintentionally is no more an action than falling off a cliff is. This indicates that possibilism is actually a bit too simplistic. There is something about abilities that sets abilities apart from physical possibilities. And whatever that feature is, it does not seem to be properly captured by the possibilist.

Thirdly, there was the problem with degrees and the corresponding kind of context sensitivity that attaches to ability statements (→ 3.5). The problem with possibilism is that the existential quantifier does not seem to be particularly well-suited to account for degrees. Once there is *one* performance world among the relevant worlds, the possibilist condition is satisfied. There does not seem to be any room for the kind of scale that would have to be established in order to account for differences of degrees that trace back to differences on the dimension of reliability – i.e. the range of circumstances in which the agent manages to deliver a performance. The ordering source does not provide help here. Degrees of abilities fail to be accounted for by the possibilist. As a consequence, the context sensitivity of ability statements that results from the gradable nature of abilities remains mysterious as well.

Fourthly, there was the formal problem with the possibilist's core claim: ability statements do not express restricted possibility statements, because ability statements do not obey the most fundamental logical laws governing possibility statements. Since the possibility operator is defined by the laws that govern it, this is a decisive problem. Possibilism is deeply flawed.

4 The success view I – Agentive abilities

So far, this book has been rather destructive. Let us now switch into construction mode. In this and the next chapter, I will take all of the insights from the last two chapters and use them to come up with a new, and, as I will argue, superior view of abilities: the success view of ability. I will proceed in two steps. In this chapter, I'll be dealing exclusively with agentive abilities and developing the success view for abilities of that kind. In the next chapter, the view is going to be extended to cover non-agentive abilities as well. By the end of the next chapter, we'll have arrived at a fully general and comprehensive view of abilities – the success view of abilities *tout court*.

Why start with agentive abilities? For three simple reasons. First, the paradigmatic cases of abilities are of the agentive kind. When asked to name a few abilities, most people come up with abilities to play instruments, do some sport or other, drive a car, cook a dish, play a game, and so on. This is no coincidence. A brief look into the Corpus of Contemporary American English (COCA) makes it evident that ascriptions of agentive abilities are by far the most prevalent ones. I take it to be reasonable to start an investigation of the subject matter from there.

Secondly, the success view of agentive abilities will turn out to be rather concrete where the success view of abilities *tout court* will necessarily have to remain somewhat abstract. In the following sense: abilities *tout court*, I will argue, are a matter of success; and success, in turn, I will argue, is a matter of a certain modal tie between certain triggers and responses. The good thing about agentive abilities is that we can actually specify what the crucial trigger is in such cases; namely an intention to ϕ (or, on a weaker reading, some other properly related intention). When it comes to the success view of abilities *tout court*, all we can say is that the trigger will have to be such that ϕ-ing in response to it counts as a success. I therefore take it to be helpful to have worked out the success view of agentive abilities before moving on to the more abstract formulation of the view. (If this sounds somewhat obscure to you at this point, fear not. All of this is going to be developed very systematically in this and the next chapter.)

Thirdly, and relatedly, the success view of agentive abilities can in large parts avail itself of descriptive terms where the success view of abilities *tout court* employs normative terminology. In the typical cases, to have an agentive ability is for a certain modal tie between the intention to ϕ and performances of ϕ-ing to obtain; no normative vocabulary here. For an agent to have an ability *tout court* is for there to be a certain modal tie between a certain trigger and the agent's ϕ-ing, where the trigger is specified as one in response to which ϕ-ing counts as a success. What counts as a success obviously has a normative dimension to it. Not that there is anything wrong with this normative component. In fact, I take it to be clear that abilities have some normative dimension to them. Yet, it is not particularly easy to pinpoint where exactly the normativity stems from. It is therefore somewhat comforting that the more specific version of the view – the success view of agentive abilities – can in large parts do away with normative notions.

All this being said, let's forget about non-agentive abilities for the time being, and concentrate on the paradigmatic instances of abilities: abilities to perform actions. Let us focus on abilities like the ability to sing, to dance, or to jump; to eat, to cook, or to pour water; to sketch a tree, to chase a cat, or to hole a putt; and so on. Let's focus exclusively on *agentive* abilities and forget about the rest.

Here is a road map for the chapter. In the next section, 4.1, I will introduce the basic framework of the success view of agentive abilities in quite some detail. As I go along, it will become evident that the success view incorporates key insights both of the simple conditional analysis and possibilism. That is because the success view of agentive abilities incorporates both the idea that abilities have something to do with a modal tie between intentions and performances and the idea that abilities are always had in view of contextually variant sets of facts. In a way, then, the success view of agentive abilities can be seen as a hybrid view of the simple conditional analysis and possibilism. I view this as a merit. I take it to be highly likely that other philosophers were not completely off course in their assessment of the topic.

In section 4.2, I am going to talk in some detail about the notion of an intention as it figures in the formulation of the view and provide a rationale for formulating the success view in terms of intentions and not some other candidate motivational state on the agent's part.

In sections 4.3–4.7, I show that the success view of agentive abilities is superior to both the simple conditional analysis and possibilism in that it accounts for all of the adequacy conditions (apart from accounting for non-agentive abilities, of course) and does not run into any of the problems that beset the other two views.

More specifically, I will argue that the success view provides an account of degrees and the corresponding kind of context sensitivity that attaches to ability statements (→ 4.3), circumvents the problem of the impeded intention (→ 4.4), provides a fully comprehensive account of general and specific abilities (→ 4.5), yields an understanding of the workings of masks (→ 4.6), and does not only avoid the formal problem that besets possibilism, but actually explains where that problem stems from (→ 4.7). In view of those merits, I will conclude, by the end of the section 4.7, that the success view is a very promising view indeed.

Sections 4.8 and 4.9, finally, are devoted to the discussion of two objections to the success view that come to mind quite naturally. In section 4.8, I'll discuss the objection that the success view runs into problems with what I will call "unintentional agentive abilities" – agentive abilities which can only be exercised as long as the agent does not intend to exercise them.

In section 4.9, the worry is refined by examining a specific kind of unintentional agentive abilities: abilities that can only be exercised as long as the agent does not intend to exercise them, because the agent is epistemically impaired in certain ways. In response to these worries, I will introduce a somewhat weaker version of the success view, which accommodates such cases. By the end of the chapter (→ 4.10), we'll be in a position to formulate a fully comprehensive account of agentive abilities.

I would like to emphasize beforehand that I am hugely indebted to Manley and Wasserman's work on dispositions. The idea that abilities have something to do with proportions is greatly inspired by their work. On their view, x has a disposition to M, "if and only if x would M in some suitable *proportion* of [stimulus cases]" (Manley & Wasserman 2008: 76, my emphasis). Although Manley and Wasserman's account of dispositions differs from my views on abilities in some crucial respects (their view is a more sophisticated version of the conditional analysis of dispositions and is going to be discussed in detail in chapter 6.2), I nevertheless view it as a very close relative to the success view of ability in that it emphasizes the notion of a proportion – a feature that can be found in a variety of views, all of which are going to be discussed in chapter 6.

4.1 The general framework

On the view I want to develop and defend in this chapter, having an agentive ability is a matter of modal success. More specifically, I will argue that for an agent to have an agentive ability to ϕ, that agent's intentions to ϕ and her performances of ϕ-ing have to match up in the right way across a certain portion of modal space. Agentive abilities are a matter of a specific modal tie between intentions and performances on this account.

As you will note, this idea pays tribute to the intuition that also drives the simple conditional analysis of abilities. According to the simple conditional analysis, an agent has an ability if and only if the agent would ϕ if she intended to ϕ. Thus, the simple conditional analysis, too, postulates a specific modal tie between the agent's intentions and her performances. But as we have seen in chapter 2, that tie is not plausibly construed in terms of a counterfactual conditional.

What is the modal tie that matters, then? Here is my suggestion. According to what I will call *the success view of agentive abilities*, or simply SUCCESS$_{AA}$, in what follows,

> SUCCESS$_{AA}$. an agent S has an agentive ability to φ if and only if S φ's in a sufficiently high proportion of the relevant possible situations in which she intends to φ.

Thus, a darts player has the ability to hit the bull's eye if and only if she hits it in a sufficiently high proportion of relevant possible situations in which she intends to hit it. An engineer has the ability to construct a super-efficient engine if and only if she succeeds in constructing one in a sufficiently high proportion of relevant possible situations in which she intends to construct one. And a pianist has the ability to play a certain piece if and only if she succeeds in playing it in a sufficiently high proportion of relevant possible situations in which she intends to play it.

In fact, this is a simplification. As we will see later on, SUCCESS$_{AA}$ is only approximately correct. Really, the proportion that matters will have to be weighted (→ 4.3) and the agent's intention does not necessarily have to be the intention to φ, but may well be some other intention, in response to which φ-ing counts as a success (→ 4.8). But those complications need not bother us right now. SUCCESS$_{AA}$ is close enough to what I take to be the truth to get along with it throughout the largest part of the chapter. The full-fledged version of the success view of agentive abilities, which does justice to the full amount of complexity of its subject matter, is going to emerge as we go along and can be found in the upshot in section 4.10. For now, let's keep things as simple as possible.

What is a proportion? A proportion, as I am using the term, is simply a portion or part in its relation to the whole.[1] The proportion of blondes among the Swedes, for instance, is the portion of blonde Swedes in relation to all Swedes. Mathematically, a proportion corresponds to a certain ratio. To use our example again, the proportion of blondes among the Swedes corresponds to the ratio of blonde Swedes to the totality of Swedes. It can be expressed by a fraction. The ratio of blonde Swedes to all Swedes is v, such that v = size of the set of blonde Swedes/size of the set of all Swedes.

---

[1] I follow the Oxford dictionary here, which defines "proportion" as "a portion or part in its relation to the whole; a comparative part, a share; sometimes simply, a portion, division, part" (Simpson & Weiner 1989).

The same goes for the proportion of performance cases among the relevant intention situations, as it figures in SUCCESS$_{AA}$. It, too, corresponds to a ratio: the ratio of relevant intention situations in which the agent ɸ's to the totality of all relevant intention situations. Again, this can be expressed by a fraction. The ratio that matters is v, such that v = size of the set of relevant intention situations in which the agent ɸ's/size of the set of the all relevant intention situations. We can call this ratio the agent's *modal success rate*. Hence, SUCCESS$_{AA}$ states that an agent has the ability to ɸ if and only if her modal success rate is high enough.

It is important to note that the modal success rate is not to be conflated with the agent's quota of successful attempts in the actual world. Let's distinguish the two ratios very carefully. The modal success rate is the proportion of the success cases among the relevant *possible* situations in which the agent intends to ɸ. The agent's quota of successful performances in the actual world, in contrast, is what we can call her *track record*[2]: it is the proportion of the success cases among the *actual* intention situations.

The agent's track record plays a heuristic role. By looking at an agent's track record, we find out about situations in which the agent's intentions were successfully realized and situations in which they weren't. That gives us good reasons to infer certain beliefs about patterns of success across possible worlds. In the end, though, it's the modal success rate that counts.

To get a better grip on SUCCESS$_{AA}$, let's visualize the view by means of a Venn diagram.

---

[2] I borrow this term from Greco (2007: 60f.)

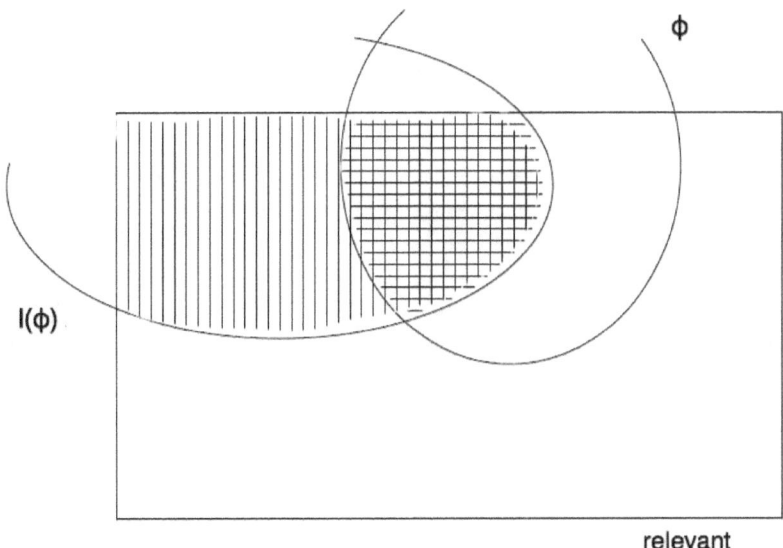

What matters for an agent's agentive ability to φ, according to SUCCESS$_{AA}$, is the proportion of φ-ing cases among the relevant intention situations. In the diagram, we have the relevant possible situations in the box. In the vertically dashed area, we have the relevant intention situations. In the horizontally dashed area, we have the relevant intention situations in which the agent φ's. What matters is the proportion of the horizontally dashed area among the vertically dashed area. If it is sufficiently high, the agent has the ability to φ. If not, she lacks it.

Quite simple thus far. And quite intuitive, I hope. Let's dive deeper. Thus far, I have only emphasized the similarities between SUCCESS$_{AA}$ and the conditional analysis; both stress the importance of a certain modal link between the agent's intention to φ and the agent's effective performances of φ-ing. But a second glance at SUCCESS$_{AA}$ reveals that the condition also draws on an important idea of possibilism. For what matters for an agent to have an ability, according to SUCCESS$_{AA}$, is the agent's performances across a certain restricted domain of possible situations. And this feature is something the view obviously shares with possibilism.

While the proponent of the simple conditional analysis always moves to the intention worlds that are overall closest to the actual world, the possibilist restricts the possible worlds in correspondence to whatever features of the actual world are of interest in a given context and evaluates the agent's performances in any of those worlds. The exact same procedure is used in the case of SUCCESS$_{AA}$:

we first determine the relevant possible situations. For this reason, I will often employ Kratzer's terminology and refer to the relevant possible situations as the modal base and to the facts that determine that modal base as the facts "in view of which" the agent has an ability.

Note that SUCCESS$_{AA}$ speaks of situations where possibilism speaks of worlds. That is because what matters, according to SUCCESS$_{AA}$, is the ratio of performances of φ-ing to intentions to φ. Worlds, of course, can contain many instances of intending to φ. When we think about the ratio of performances to intentions, all of those instances count. That is why worlds are not fine-grained enough to figure in SUCCESS$_{AA}$. For this reason, SUCCESS$_{AA}$ is formulated in terms of situations. A situation, as I use the term, is a proper part of a world. It is a world at a certain stretch of time.

Which situations go into the modal base will vary. Like possibilism, SUCCESS$_{AA}$ is a contextualist view about abilities in the sense that the set of the relevant possible situations varies across contexts. Let's make this explicit. SUCCESS$_{AA}$, as I will defend it in the course of this book, is to be understood as the view that

> SUCCESS$_{AA\_context}$. an agent S has an agentive ability to φ if and only if S φ's in a sufficiently high proportion of the relevant possible situations in which she intends to φ, *where the set of the relevant possible situations will vary across ascriber contexts.*

Both possibilism and SUCCESS$_{AA}$ restrict the possible worlds and situations, respectively, to the ones that count as relevant in a given context, then. In both cases, certain facts of the actual world are held fixed across modal space. Others vary wildly. Sometimes, any situation will count as relevant in which the agent's intrinsic features are as in actuality. Sometimes, to count as relevant, a situation will additionally have to resemble the actual situation with respect to varying sets of extrinsic features of the agent.

In connection with possibilism, we have seen that this contextual restriction to different sets of worlds in terms of relevance took us quite a few steps towards an account of more or less general or specific abilities. As we'll see later on in some detail (→ 4.5), SUCCESS$_{AA}$ preserves this extremely fruitful feature of possibilism, but moreover provides the resources for a full-fledged account of general and specific abilities.

There are two important differences between SUCCESS$_{AA}$ and possibilism. First, in contrast to the possibilist, we do not look at the agent's performances across *all* relevant possible situations, but only across a certain subset of those situations. The link between intentions and performances, which is also what proponents of the simple conditional analysis are after, is established by adding a second restriction on the situations that count. What matters is not just the agent's performance across the relevant possible situations in general, but rather across the relevant possible situations in which the agent intends to ɸ. This second restriction, which is clearly not part of the possibilist framework, is what secures the modal link between intentions and performance. But it does so by a core means of possibilism – by imposing restrictions on the modal realm.

The second difference is that according to SUCCESS$_{AA}$, the possibilist is wrong in assuming that abilities can be analyzed in terms of the possibility operator. While abilities are indeed a matter of the agent's performance across relevant possible situations, one performance case is rarely enough. That is why possibilism ran into the destructive formal problems we looked at and had trouble accounting for general abilities and degrees. Possibilists get the modal force of ability statements wrong; ability statements are not possibility statements.

What is the right modal force? On my view, abilities have to be understood in terms of what we can call "proportional quantification": there has to be not one, but a sufficient *proportion* of relevant intention situations in which the agent ɸ's.

What does "a *sufficiently high* proportion" amount to? That will vary – and rightly so. We know that there is context sensitivity involved in the degree of an ability that is required for an ability ascription *simpliciter* to apply. And as I will lay out in due course (→ 4.3), the (weighted) proportion of performance cases among the relevant intention situations corresponds to the degree of an agent's ability. More specifically, I will argue that the higher the (weighted) proportion of performance cases is among the relevant intention situations, the higher the agent's degree of ability. And so, since it varies which degree of ability is required for the agent to count as having the ability *simpliciter* in a given context, it will also vary which proportion of performance cases counts as sufficient.

Despite some important similarities, then, SUCCESS$_{AA}$ differs substantially from possibilism. What remains to be done is to set the view apart from the simple conditional analysis a bit more carefully as well. The major difference between the success view and the simple conditional analysis is that the success view is not formulated in terms of a counterfactual conditional. Now that we have a grip on the idea of proportional quantification and the restriction to relevant possible situations that figures so prominently in SUCCESS, it should be easy to see that this difference has far-reaching consequences.

By formulating the simple conditional analysis in terms of a counterfactual, the proponent abilities are analyzed in terms of the agent's performance in the closest worlds in which the agent intends to ɸ. An agent has an ability to ɸ, on that view, if and only if the agent ɸ's in the closest worlds in which she intends to ɸ. SUCCESS$_{AA}$ is at the same time more liberal and more restrictive than that. It is more liberal since the agent need not φ in the closest intention situations. A φ-ing performance in one of the less close, but still relevant intention situations is just as valuable as in the very closest ones. At the same time, SUCCESS$_{AA}$ is more restrictive, because in most contexts one performance case just won't do, no matter how close it is: the agent has to φ in a sufficient proportion of the relevant intention situations, and as we will see later on (→ 4.3; 4.5), that will usually require that she φ's in more than just one situation.

The main idea of SUCCESS$_{AA}$ should have come across. But there are many details which are going to be provided as we go along. In the next section (→ 4.2), I'll talk about the notion of an intention and provide a rationale for formulating the view in terms of that motivational state and not another. Afterwards, in sections 4.3 – 4.7, we'll see that the success view is very powerful. Rightly developed, it delivers a fully comprehensive account of agentive abilities, which accounts for all of the adequacy conditions for a comprehensive view of such abilities and does not run into any of the problems that beset the simple conditional analysis and possibilism, respectively.

4.2 The proper motivational state

The success view shares one important feature with the simple conditional analysis: both stress the importance of a certain modal link between the agent's intentions to φ and the agent's effective performances of φ-ing. When I introduced the simple conditional analysis, I said that different versions of the view differ with respect to the motivational state that figures in the antecedent of the counterfactual. Different versions of the view feature motivational states as diverse as trying, intending, wanting, choosing, and desiring. For reasons of exposition, I did not dwell on those differences, but focused on the similarities between the various versions instead. For plasticity, I just picked one motivational state and formulated the simple conditional analysis in terms of intentions. But I could just as well have formulated it very broadly as the view that an agent has an ability to φ if and only if the agent would φ, if the agent were *properly motivated* to φ.

I could have availed myself of the same non-committal formulation in my formulation of SUCCESS$_{AA}$. SUCCESS$_{AA}$ could have been formulated as the view that an agent has an agentive ability to φ if and only if the agent φ's in a sufficiently high proportion of relevant possible situations in which the agent is properly motivated. Instead, I have chosen a more committal formulation: SUCCESS$_{AA}$, as I have formulated it, postulates a modal tie between a particular kind of motivational state and the agent's performances. What counts, on my view, is the tie between the agent's *intention* to φ and her effectively φ-ing. In this section, I will elaborate a bit on the notion of an intention, as it figures in the analysis, and on the rationale for focusing on intentions and not some other candidate motivational state.

An intention, as I am using the term in the course of this book, is simply an *action-initiating* propositional attitude in the sense that it is part of its causal role that it will typically initiate behavioral episodes corresponding to its content. I will assume that the behavioral episodes which are caused by intentions (in the right way)[3] are actions. Further, I will assume that it is the causal impact of an intention that explains the intentionality of an action. I hope to be firmly ensconced in the action-theoretical common sense or at least the respectable tradition with these assumptions.

We can differentiate between intentions to φ right away and intentions to φ in the future. Intentions, as they figure in SUCCESS$_{AA}$ are intentions to φ right away. Intentions to φ in the future can be given up, and it therefore does not count against an agent's ability, if an agent intends to φ in the future, but does not φ when the time comes. So "intention", as I use it, is always the intention to φ at t, where t is, roughly, the next moment.

Let me now talk about the rationale for focusing on intentions, so understood, and not some other candidate motivational state. First, the motivational state figuring in the account has to be action initiating the sense that it will typically lead to action. A state that is motivational but not action-initiating in this strong sense will not do. To see that, take ordinary desires. Desires are motivational, but not action initiating. They constitute, *qua* desire, a mere *prima facie* motivation towards the desired action. This has two reasons.

On the one hand, desires can be overridden by other desires. I can have a desire to go dancing, but that desire can be overridden by an even stronger desire to go to sleep (yes, I am getting old). In that case, dancing will not be the favored course of action. Instead, I'll presumably end up curled up in my bed. On the other hand, desires alone will not do anyway; without a proper belief, no action will ensue. Even an overriding desire to go dancing will only result in dancing, if I believe that there is a party somewhere. Otherwise, I'll end up sleeping all the same.

These features of desires indicate that desires are not the suitable motivational states to figure in an account of abilities. My not acting on a desire to

---

[3] "In the right way" means "by non-deviant causal chains". How to think of non-deviance is a problem of action theory and falls outside the scope of this book. For an overview see (Wilson & Shpall 2012, section 2).

go dancing, which is either overridden by an opposing desire or paired with an action-inhibiting belief, does not speak against my ability to go dancing. That is why the success view cannot be formulated in terms of desires or any other merely *prima facie* motivating state, such as wants. We get distorted results, if we evaluate the proportion of dancing cases among the relevant possible situations in which the agent is merely *prima facie* motivated to dance. Any situation in which the agent has an overriding desire or want to go to sleep or believes that there is no party around, would have to be counted against the agent's ability to dance. Certainly, this is the wrong outcome. What we need is an action-initiating state: a state which typically initiates action. Intentions are just that.

Yet, the need for an action-initiating state does not rule out a number of different motivational states. It does not speak against a formulation of the success view in terms of choices, decisions, tryings, or attempts, for instance. To see why those motivational states are ill-suited to figure in the analysis, we need to focus on a different requirement that goes along with a satisfactory formulation of an account of agentive abilities: the motivational state had better itself not be an action.

Otherwise, a regress looms. The success view, as it stands, is supposed to offer an account of *agentive* abilities – abilities to perform actions. The way the view aims at this goal is by postulating a modal tie between a suitable motivational state to perform some action on the one hand and the effective performance of the action on the other. If this is true, however, then the motivational state figuring in that modal tie had better not itself be an action. Otherwise, one will rightly wonder what it is for the agent to have an ability to perform *that* kind of action.

For illustration, let's assume that the success view were formulated as the view that an agent has an agentive ability if and only if the agent φ's in a sufficiently high proportion of relevant possible situations in which the agent *tries* to φ, and let's assume, as many have, that trying is an action.[4] Then the question will naturally arise what it is for an agent to have the ability to perform *that* action: what is it for an agent to have the ability to try?

For the proponent of the success view, the only reasonable available answer is to state that an agent has an ability to try if and only if the agent tries in a sufficiently high proportion of the relevant possible situations in which the agent *tries to try*. And so on. *Ad infinitum*, with ever less plausibility.[5] The same goes for choosing, deciding, attempting, and any other motivational

---

4 See for instance Adams (1994).
5 Note that Moore (1912, ch. 7) does not seem to think that there is any problem with this.

state that seems to be an action itself. We can conclude that the success view had better be formulated in terms of a motivational state that is itself clearly not an action. Intentions, I take it, meet this requirement.

4.3 An account of degrees and context sensitivity

In this and the next four sections, I will, piece by piece, demonstrate the merits of SUCCESS$_{AA}$. In this section, I will turn to the way the success view accommodates degrees of abilities and the corresponding kind of context sensitivity that attaches to ability statements. Providing an understanding of those phenomena is a central task in developing a viable account of abilities. In fact, it is one of the adequacy conditions for a comprehensive view of abilities. As I have laid out in chapter 1.3, any comprehensive view of abilities will have to accommodate degrees and the corresponding kind of context sensitivity of ability statements.

On the view I will lay out in this section, the crucial tool provided by SUCCESS$_{AA}$ is the notion of the agent's modal success rate. Roughly, we can say that the higher the proportion of ϕ-ing cases is among the relevant intention situations, the higher the degree of an agent's ability.[6]

Let's recall that abilities can be graded along two dimensions. On the one hand, there is the dimension of reliability. It can count, towards the degree of an agent's ability, how large the range of the circumstances is in which the agent manages to perform an action. On the other hand, there is the dimension of achievement. It can count, towards establishing the agent's ability, how high the quality of the agent's performance of an act is that the agent manages to deliver.

In what follows, I'll give an account of both dimensions individually, but on top of that I'll also sketch an account of agents' degrees of abilities quite generally. That means that I'll give an account of degrees which holds, no matter which dimension carries which weight. It thus unifies the two dimensions and reveals a way to think of a mechanism of how to set them off against each other in contexts in which they are set off.

The account of degrees of reliability is quite easily developed. SUCCESS$_{AA}$ is tailor-made for the understanding of degrees along that scale. Reliability has to

---

[6] A very similar account is developed for degrees of dispositions by Manley & Wasserman (2008). Manley and Wasserman's account is a very sophisticated version of the conditional analysis, but it also features the notion of proportions of (centered) worlds and uses that notion to account for degrees. See also Maier (2013). Both accounts are going to be discussed in detail later on (→ 6.2, 6.4).

do with the range of possible circumstances in which an agent manages to perform a certain action. The modal success rate measures just that: the range of the possible circumstances in which the agent's intentions are followed by the corresponding action. Thus, the modal success rate *is* the measure for reliability. The more reliable archer is one who hits the target in a higher proportion of relevant possible situations in which she intends to hit it. The more reliable weather forecaster is one who predicts the weather correctly in a higher proportion of relevant possible situations in which she intends to predict the weather. And so forth.

At first sight, it is not obvious that SUCCESS$_{AA}$ accounts equally smoothly for the dimension of achievement. To see the issue, think of the two pianists, one of whom plays perfectly on every occasion, while the other one plays poorly on every occasion. On the dimension of achievement, the first pianist is better. But *prima facie*, this dimension does not fit well with the success view. The notion of being better in terms of achievement seems to be essentially qualitative. It does not seem to have to do with the range of cases in which one performs at all. What counts is how *well* one performs. And that, it might seem, cannot be accounted for in terms of the purely quantitative measure that is given by comparing proportions of cases in which performance takes place. In a nutshell: the modal success rate in ϕ-ing does not seem to be qualitative enough to account for the dimension of achievement.

Fortunately, this is less of a problem than it may seem at first sight. Proportions matter. But not every ϕ-ing case is counted equally when it goes into the proportion of ϕ-ing cases among the relevant intention situations. Instead, we need to *weigh* the cases we are counting.[7] Here is how.

The qualitative dimension that enters in on the level of achievement can be modeled quantitatively by giving certain situations a higher value the closer the performance in those cases comes to some (contextually determined) ideal. The idea behind this is simple. If someone plays a piano piece flawlessly, that counts more towards her ability to play the piano than her playing the same piece badly. If someone plays a piece with verve and emotion, that counts more towards her ability to play the piano than if she plays the same piece like an automaton. And so forth. This notion of "counting towards the ability" can be captured by adding extra value to a situation the more the performance in that situation counts towards the ability in question. The more verve an agent's play exhibits in a given

---

[7] This is also suggested by Manley & Wasserman (2008) in connection with the gradability of dispositions. In assessing the fragility of a vase, it is not only important in which proportion of stimulus worlds it breaks, but also how extensively it breaks. Apparently, shattering counts more towards an object's fragility than cracking.

case, the higher the value we attach to that case. The less mistakes an agent makes in a given case, the higher the value we attach to those cases.

Depending on the context, it will vary which qualities matter most. In some contexts, flawlessness is the major virtue we are after. In others, verve is more important. This is mirrored in the values we attach to the cases. The more important verve is, the more value we attach to the verve-cases. The more important flawlessness is, the more value is given to the flawlessness-cases. And so on. We can think of that in terms of numbers. An ideal performance gets a 1. A performance that is too miserable to count as an act of ϕ-ing at all gets a 0.

When is an agent better on the achievement level? That depends. Sometimes, all that matters is her best performance across the relevant intention situations. When we are evaluating two piano players' abilities to play a certain piece, we may sometimes be willing to count that agent more able on the achievement level who manages to deliver the best performance, no matter how miserably the agent plays in most other cases. In such cases, the highest value wins, so to speak.

Perhaps there are also contexts in which the lowest achievement value is what counts. It is not implausible, for instance, that there are contexts in which an emergency doctor's ability to prevent damage is counted as higher the fewer people die under her care and not the more people come out completely healed in the end. In such contexts, the higher the lowest value of cases in which damage is prevented, the better the doctor on the achievement level.

Finally, there also seem to be cases in which we are interested in the average value of the agent's performances. Usually, a hedge fund manager who plays full-risk all the time and wins a fortune every once in a while losing huge amounts of money in most cases does not count as better, in terms of achievement, than someone who plays it cool and wins a few percentages in each transaction. Here, what often counts seems to be the average achievement value.

Which achievement value is of interest in a context can vary and often we will have to set the dimensions off against one another. A very high peak value will sometimes outweigh a poor average and a devastating off-peak value. But likewise, a very low off-peak can outweigh a good average value and an amazingly high peak value. Let's assume, for the time being, that the average value is always what counts. I'll come back to the interplay between the various achievement values in due course.

Let's take stock. What we have is an account of both degrees of reliability and degrees of achievement. An ability's degree of reliability corresponds to the proportion of relevant intention situations in which the agent performs the intended action. An ability's degree of achievement corresponds to whatever achievement value matters to us, but for the time being we assume it corresponds to the average achievement value of the agent's performances. Sometimes, we are interested in reliability only. Sometimes, in achievement only. Which of the two matters varies with the kind of ability we are considering, and presumably also with other context factors. So far, so good.

In many cases, however, this is not enough. Very often – in fact: usually – we are not just interested in one of the two dimensions. Usually, an agent's overall degree of an ability depends on both her degree of reliability and her degree of achievement. What we really want, then, is a unified account of degrees, taking into account both the dimension of reliability and the dimension of achievement at once. We want what I would like to call a *weighted proportion*[8] – a value that is higher the better someone performs in the more cases.

Obviously, there are various ways of generating such a value, but not all of them work equally well. For instance, we will have to make sure that reliability and achievement as a whole can be of unequal importance for the degree of an agent's ability. Sometimes, reliability is what matters most. Sometimes, achievement.

Here is a toy model for the way in which the two dimensions might be set off against each other. In a first step, we evaluate the dimension of reliability. We determine the proportion of ϕ-ing cases among the relevant intention situations. The higher the proportion, the higher the number. In a second step, we determine value for the agent's achievement. For simplicity, we are working with the average value, and that value is obtained by adding the values for the agent's individual performances and dividing it by the number of relevant intention situations. The situations in which she does not perform at all get a 0.

We can now add these values. But before we actually add them, we have to determine how much weight to put on each value. We can do this by using a multiplier. Let m and n be multiplier variables. Let R be the value for reliability. Let A be the value for the agent's achievement. Then the agent's success rate, SR, will be determined along the following lines:

$$SR = (m \times R) + (n \times A)$$

If we are interested in reliability only, we set n = 0. If we are interested in achievement only, we set m = 0. If we are interested in both, we weigh them by attaching to each a multiplier that suits our focus. By that procedure, we obtain the weight-

---

8 I borrow this term from Manley & Wasserman (2008). Manley and Wasserman give no account of how the correlates of reliability and achievement are to be set off in the case of dispositions, but as far as I can see, they, too, will need something like the toy model I will suggest for degrees of abilities.

ed proportion we were after – a value that is higher the better the agent performs in the more situations and in which the dimensions of reliability and achievement can be set off against each other according to the focus we put on each dimension in a given context.

By the same kind of procedure, we can determine the agent's overall achievement value, by the way. The challenge, recall, was basically the same: the agent's achievement often depends on a variety of values – the peak value, the off-peak value, and the average value of the agent's performance. Obviously, this interplay can be modeled along the same lines as the interplay between achievement and reliability as a whole. We can simply add all of the values which count and determine the extent to which they go into the sum by attaching appropriate multipliers to each of them beforehand. Whatever overall value comes out for the agent's achievement is the one we are inserting into the formula for the agent\s overall success rate in ϕ-ing.

I should emphasize that I am not arguing that we really do this kind of calculation when we evaluate an agent's degree of ability. What I have described as a calculation procedure is not really a procedure that falls into distinct steps. In fact, it is not even a procedure at all. It is a toy model of degrees of abilities in a possible worlds framework. It gives us an idea of how the notion of degrees of abilities, though highly qualitative in that it involves how highly we appreciate a given performance in a given context, can nevertheless be modeled in quantitative terms: as a numeric measure we project onto possible situations and which will change across contexts, depending on the one hand on our standards for what counts as a performance of ϕ at all, what counts as a good or bad performance, and what counts as a brilliant performance, and on the other hand on how important reliability and achievement are, respectively.

In virtue of the fact that the model is for illustrative purposes only, it is not of crucial importance that you agree with every detail of it. Perhaps you think there is more to be said about the dimension of achievement; perhaps you think that my discussion of the exemplary cases is not fully satisfactory; perhaps you have other quarrels with the details of the account. That is okay. As long as those quarrels do not concern the structural ingredients of the account (i.e. an account of degrees in terms of weighted proportions), I am fine with that. The ways in which we grade abilities are obviously very complex, and my account in terms of the two dimensions and the ways in which they are set off against each other is perhaps too coarse to account for every single way in which two agents' degrees of ability can vary. So be it, then.

What matters is that the basic notion of the success view, the notion of a success rate, is structurally suited to account even for the qualitative parts of our thinking about degrees and can account for the fact that various different dimensions, such as achievement and reliability, go into our calculations when we evaluate an agent's overall degree of ability. Proportions matter – all we have to do is weigh the cases that are counted in an appropriate way. What I have outlined is one structural idea of how to do so. There may be other ways, and there may definitely be ways to improve the model I have sketched. I am content if my understanding of degrees is *structurally* on the right tracks, and I believe it is.

Here is the upshot of this section. According to the success view of ability, an agent has the ability to φ only if the agent φ's in a sufficiently high proportion of relevant possible situations in which she intends to φ. In the preceding paragraphs, I have elaborated on the right understanding of "proportion". What matters is not just the sheer number of φ-ing cases. Rather, "proportion", as it occurs in the formulation of the success view, has to be interpreted as the weighted proportion that corresponds to the weighed sum of both, the agent's reliability value and her achievement value. The weighing is a matter of context. Sometimes, reliability counts more. Sometimes, achievement. The success view accounts for that.

Context sensitivity costs nothing on this picture. Once we have the scale established by varying success rates, and hence varying degrees of ability, the context simply sets a threshold on that scale. The threshold is provided by the notion of sufficiency. For an agent to have an ability *simpliciter* in a given context, the agent has to exhibit a sufficiently high weighted success rate in φ-ing. And which weighted success rate counts as sufficient will vary. In what follows, I will not speak of the weighted proportion and success rate anymore. I take it to be clear that "proportion" and "success rate" have to be understood along those lines, and I trust that the reader will bear it in mind.

4.4 Impeded intentions and the existential requirement

SUCCESS$_{AA}$ analyzes abilities in terms of a modal tie between intentions and performances. In this respect, it resembles the simple conditional analysis rather closely. For this reason, one very pressing question is whether SUCCESS$_{AA}$ may not actually fall prey to the problem of the impeded intention as well. In view of the fact that the postulation of the counterfactual tie gave rise to the problem of the impeded intention in the case of the simple conditional analysis, the postulation of a very similar tie may seem like a very bad idea.

To evoke the problem, recall that the counterfactual tie between the agent's intentions and performances postulated by the simple conditional analysis is obviously not sufficient for an agent to have an ability. Reconsider Betty, our coma patient. For Betty, the counterfactual comes out true: she would raise her arm if she intended to raise it. That is because the closest worlds in which she intends to raise her arm are worlds in which she is not in a coma. And in those worlds, she would presumably also raise her arm.

An analogous worry seems to apply to $SUCCESS_{AA}$. For may not the modal success rate be quite high in the case of a coma patient? The agent's intention to raise her arm will only occur in situations in which the agent is not in a coma, after all. And in such situations, nothing impedes her raising her arm. Hence, it seems as though the modal success rate will be averagely high for the coma patient. If this were true, then $SUCCESS_{AA}$ would run into an analogous problem as the simple conditional analysis: the modal success rate is averagely high, while the agent lacks the ability, because she cannot form the intention to begin with. Like the simple conditional analysis, the condition stated in $SUCCESS_{AA}$ would turn out to be insufficient in that case.

Luckily, this objection relies on a misunderstanding of $SUCCESS_{AA}$. $SUCCESS_{AA}$ states that to have an agentive ability to φ, the agent has to φ in a sufficiently high proportion of *relevant possible situations* in which she intends to φ. *This* condition is not met in the case of the coma patient. In contexts in which we want to say that the coma patient lacks the ability to raise her arm, the set of relevant possible situations will contain only situations in which the agent is in a coma at the outset. But among *those* situations, there will be no intention cases to begin with. And from this it follows that there will not be a sufficiently high proportion of performance cases among the relevant intention situations. Let me explain.

When there are no relevant intention situations, to determine the proportion of performance cases among the relevant intention situations, one would have to determine the proportion of performance cases among the empty set. And depending on the exact mathematical understanding of "proportion", that proportion will either be unspecified or zero. Either way, it will not be the case that there is a *sufficient* proportion of performance cases among the relevant intention situations. $SUCCESS_{AA}$ fails to be met either way, then.

Let me elaborate on this last point in some more detail. First, let's recall that $SUCCESS_{AA}$ applies proportional quantification to the relevant intention situations. What $SUCCESS_{AA}$ states is that an agent has an ability if and only if *there is a sufficiently high proportion* of performance cases among the relevant intention situations. This condition can be violated in two ways. It is violated, first, if there is no proportion of performance cases among the relevant intention situations to begin with. It is violated, secondly, if there is a proportion, but the proportion is not sufficiently high. Depending on how exactly we spell out the notion of a proportion, $SUCCESS_{AA}$ will fail to be met on either one of those two grounds.

If "proportion" is spelled out as a ratio, as I have done in terms of the modal success rate, then the proportion of performance cases among the relevant intention situations will be unspecified, if there are no relevant intention situations to begin with. We are dividing by zero in that case. To see that, let I be a value representing the size of the set of the relevant possible situations in which S intends to φ. Let P be a value representing the size of the set of the performance cases among those intention situations. Given the interpretation of a proportion in terms of a ratio, $SUCCESS_{AA}$ states that an agent has an ability to φ if and only if

$SUCCESS_{AA\_ratio}$. there is a value v: v = P/I, such that v is sufficiently high.

From this, it follows that the proportion of performance cases among the relevant intention situations is unspecified in the coma case. In the coma case, the value for I is zero. It will thus not be true that there is a sufficient proportion of performance cases among the relevant intention situations. There is no such proportion. $SUCCESS_{AA}$ is therefore not met in this case.

If "proportion" is given a less formal interpretation – an interpretation along the lines of a share or a part, say – then one may be able to make sense of the notion of a proportion of cases among the empty set. Even so, however, that proportion can only ever be zero in that case. No subset of the empty set will ever contain anything. If that is true, then $SUCCESS_{AA}$ will fail to be met on that interpretation of "proportion" as well, since the proportion of performance cases among the relevant intention situations will not be sufficiently high in that case. Zero situations just won't do.

We have just revealed an important entailment of $SUCCESS_{AA}$. On any sensible interpretation of $SUCCESS_{AA}$, $SUCCESS_{AA}$ is only met if there is a relevant intention situation in the first place. In other words, the view entails that it has to be possible, in the properly restricted sense, for the agent to form the intention. Call this the *existential requirement*:

EX. An agent has an ability only if there is a relevant possible intention situation.

That $SUCCESS_{AA}$ entails the existential requirement is a very welcome consequence. When we looked at the simple conditional analysis, we saw that a very natural way of responding to the problem of the impeded intention is to try and supplement the counterfactual by a further requirement. Not only does it have to be true that the agent would φ, if she intended to φ, she also has to be *able to form the intention* in the first place. The problem was that the proponent of the simple conditional analysis did not have the resources to explain what on earth the truth conditions for this supplementary claim were supposed to be.

The success view, in contrast, can do justice to that insightful intuition. It is true that for an agent to be able to ɸ, the agent has to be able to form an intention to ɸ to begin with. And what is it for the agent to be able to form the intention? Well, as a minimal requirement, it is for the existential requirement to be true: there has to be a relevant possible situation in which the agent intends to ɸ.

When the intention is impeded, the existential requirement is not met. Holding the coma fixed, there are no relevant intention situations to begin with. Hence, the proportion of hand-raising cases among the relevant intention situations is either unspecified (if a proportion is a ratio) or nil (on a less formal reading). Hence, SUCCESS fails to be met either way in such cases.

The same goes for the brainwashed follower of a cult. The agent cannot leave the cult, because she cannot form the intention to leave in the first place. SUCCESS yields exactly that. For the brainwashed agent, there is no sufficient proportion of situations in which the agent leaves the cult among the relevant possible situations in which she intends to do so. That is because there are no relevant possible situations in which she intends to leave the cult to begin with. The relevant possible situations will be situations in which the person is as brainwashed as she is. And since the brainwash prevents the person from forming the intention to leave, there will not be any situations among the relevant ones in which the agent intends to leave. The set of the relevant intention situations is therefore empty, and hence the proportion of the situations in which the agent leaves among the relevant intention situations is either unspecified or nil. The success view therefore rightly yields that she lacks the ability to leave the cult. Cases of an impeded motivation can be accounted for very smoothly, then.

Proponents of the simple conditional analysis who have followed my argument up to this point will probably perk up their ears right now. Doesn't an analogous treatment of cases of an impeded intention lend itself to them as well? Why can't proponents of the simple conditional analysis simply say that for an agent to have an ability, two requirements have to be met. First, it has to be the case that the agent would ɸ, if she intended to ɸ. Secondly, there have to be worlds in which the agent intends to ɸ and those worlds have to be similar enough.

That, it seems, solves the problem of the impeded intention. For while the counterfactual would still be met by the coma patient, the additional condition would not. Worlds in which the coma patient intends to raise her arm are simply too far out, given that the coma patient is actually in a coma. That, the proponent of the now somewhat less simple conditional analysis could say, is why the coma patient lacks the ability to raise her arm.

The account I have just sketched has actually been formulated in the literature. On Peacocke's (1999) account, S has the ability to ϕ if and only if the counterfactual is true of S and the world in which S tries to ϕ is close (309). Peacocke then spells out closeness in terms of what we can reasonably rely on. A world is close, on his account, if and only if we cannot reasonably rely on it not obtaining (ibid: 310). As it seems, then, the proponent of the simple conditional analysis can avail herself of roughly the same solution for the problem of the impeded intention, but the account would have to be extended considerably.

4.5 An account of general and specific abilities

In section 4.3, I have used $SUCCESS_{AA}$ to come up with an account of degrees of abilities and the corresponding context sensitivity of ability statements. In doing so, I have shown how one very important adequacy condition that spelled trouble for both the simple conditional analysis and possibilism is met by $SUCCESS_{AA}$. Let's now move on to the other adequacy condition. Any comprehensive view of abilities has to shed light on the distinction between general and specific abilities. Let's see what $SUCCESS_{AA}$ has to offer in this regard.

According to $SUCCESS_{AA}$, an agent has an agentive ability to ϕ if and only if she ϕ's in a sufficient proportion of relevant intention situations. The important notion here is that of relevance. Which situations count as relevant in a given context varies. And these variances generate abilities that are sometimes general, and sometimes specific.

In the case of general abilities, we are interested in what an agent can do, independently of her current situation. Hence, we are interested in the agent's abilities in view of her stable, mostly intrinsic features. In such a case, the relevant possible situations will comprise any situation that resembles actuality with respect to those features. The temporary, extrinsic features of the agent's current situation are varied. In the case of specific abilities, in contrast, we are interested in what the agent can do in view of her very situation. Hence, the relevant possible situations will comprise only situations that are like actuality with respect to the features of the agent's current circumstances. Thus, we will also hold a large number of the temporary, highly extrinsic features of the agent fixed.

For illustration, consider Fred's ability to do a handstand. Does Fred have that ability when his arm is broken? That depends. In one sense, he does – he has had training, he has often done a handstand in the past, he rarely failed when he tried to do it. But in another sense, he cannot do a handstand. He can't do a handstand here and now, broken arm and all. In the terminology of general and specific abilities, this can be expressed by saying that Fred has the general ability to do a handstand, but not the specific ability.

SUCCESS$_{AA}$ can account for all this. That is because the relevant possible situations will differ, depending on whether we are thinking about the general or the specific ability. When we are thinking about Fred's general ability, we are interested in what he can do in view of his muscular constitution and his sense of balance (plus background assumptions about the laws of nature, normality of the world, etc.). Thus, we abstract away from the broken arm and count all situations as relevant in which Fred resembles his actual self in terms of his sense of balance and his muscular constitution, say. Among those situations, there is a sufficient proportion of cases in which Fred does a handstand. SUCCESS$_{AA}$ therefore rightly yields that he has the general ability.

When we are interested in Fred's specific ability, we are interested in what he can do in view of his sense of balance, his muscular constitution *and* the fact that he has a broken arm. (Again, add the background assumptions. I won't mention that anymore as I go along.) Thus, we factor the broken arm in – only situations will count as relevant in which Fred's arm is as in actuality. Among those situations, Fred will not do a handstand in a sufficient proportion of intention situations. Hence, SUCCESS$_{AA}$ rightly yields that Fred lacks the specific ability.

We can note, that SUCCESS$_{AA}$ is superior to the simple conditional analysis and possibilism when it comes to general and specific abilities. Unlike both views, SUCCESS$_{AA}$ accounts for the fact that general and specific abilities can come apart and that we are often willing to ascribe the general, but not the specific ability to φ to an agent. Moreover, it yields a unified account of ability statements, with general and specific abilities turning out to be essentially of the same kind. In contrast to both the simple conditional analysis and possibilism it thus has the advantage of meeting the second adequacy condition for a comprehensive account of abilities. It makes sense of the distinction between general and specific abilities and it elucidates their relation.

You will have noted that the account I just suggested avails itself of the very same mechanism the possibilist suggested for the treatment of general and specific abilities. According to the possibilist, recall, the difference between general and specific abilities is a difference in the sets of facts that go into the modal base. While we factor in lots of features of the agent's actual situation in the case of specific abilities, we only hold certain stable, mostly intrinsic features fixed in the case of general abilities. With respect to the agent with the broken arm, say, we hold the broken arm fixed when thinking about what the agent can do here and now, broken arm and all, but vary the fact that the agent has a broken arm when thinking about what the agent can do quite generally, independently of her current situation. Thus, the set of the relevant possible situations is going to vary, depending on whether we are interested in an agent's general or specific ability.

With respect to these features, my own account of general and specific abilities does not differ much from the possibilist's. I draw heavily on Kratzer's idea that the difference between the two kinds of abilities is essentially one of different modal bases. Yet, there is a very important difference, having to do with the kind of quantification that is applied to the relevant possible situations or worlds.

True to her name, the possibilist applies existential quantification to the possible worlds: to have an ability is for the agent to ϕ in one of the relevant worlds. According to SUCCESS$_{AA}$, in contrast, we need to apply what I called proportional quantification: to have an ability, on this account, is for there to be a sufficient proportion of ϕ-ing cases among the relevant possible situations in which the agent intends to ϕ.

This difference matters a lot. For as we saw back in chapter 3.3, the possibilist eventually fails to account for general abilities. No matter how unrestrictive we are about the relevant worlds – no matter, that is, how few features of the actual world we hold fixed – for an agent to have a general ability to do a handstand, say, it is not enough for that agent to do a handstand in just one of the relevant worlds. For an agent to have a general ability to do pretty much anything whatsoever we expect her to show a somewhat reliable performance across various situations in those worlds.

Recall the characterizations of general abilities from several sources in the literature we looked at in the beginning in chapter 1.2. According to Whittle, a general ability is an ability to perform across a range of particular occasions. According to Honoré, a typist's claim to have the general ability to type amounts to the claim that "[s]he *normally* succeeds (…) when [s]he tries" (Honoré 1964: 465, my emphasis). And according to Maier, one's general ability "depends on one's relation to A in a wide range of circumstances" (Maier 2013: 11). This is where the kind of quantification matters. A single world just won't do. That is why according to SUCCESS$_{AA}$, an agent has an ability only if the agent performs in a sufficiently high proportion of the relevant possible situations. How large a proportion has to be in order to be sufficient can vary. But in the case of general abilities, one world will rarely do.

Thus, SUCCESS$_{AA}$ has two instead of just one parameter to account for the distinction between general and specific abilities. Not only does it vary which situations go into the modal base in each case, it also varies how large the proportion of performance cases has to be in order for the agent to have the relevant kind of ability. In part, the success view draws heavily on the tool box of the possibilist, then. But it adds one crucial tool to the box. Only thanks to that additional tool is the success view capable of offering a complete account of general and specific abilities, whereas possibilism is bound to fail for structural reasons.

So far, I have given you the main outlines of the way I am thinking about general and specific abilities. Let me now provide you with some details. I will do so by discussing some of the ideas about general and specific abilities that have been floating around rather unsystematically in the literature.

4.5.1 General abilities

The most common way in which general abilities are distinguished from specific abilities is that agents have general abilities in virtue of their intrinsic properties alone, whereas they have specific abilities in virtue of intrinsic and extrinsic properties. Berofsky, for instance, writes that "[a]n individual possessed of a power (sic!) can take advantage of an opportunity by virtue of some intrinsic characteristic" (Berofsky 2002: 197). Basically the same point is made by Vetter, who holds that

> the general ability to serve is possessed by an agent if and only if that agent has certain intrinsic features that enable her to serve; it does not depend on the agent's external circumstances. Her specific ability, on the contrary, is gained and lost with changes in her external circumstances (...). Her general ability to serve is an intrinsic property of the agent; her specific ability is an extrinsic property. (Vetter ms.)

In this passage, Vetter suggests that the generality of an ability and its intrinsicality are two sides of the same coin. Having a general ability *is* for the agent to have some set of intrinsic properties. At first sight, this seems very reasonable, given the core idea of generality we identified: a general ability is an ability that is independent of the agent's specific situation. And if it is independent of the agent's specific situation, then it has to depend on intrinsic properties of the agent alone. *Quod erat demonstrandum*. A close tie between generality and intrinsicality also seems reasonable in view of some classic examples: the ability to serve depends solely on the way the agent is constituted. The ability to walk, sing, or cook as well.

Yet, I submit that intrinsicality is not the crucial feature when it comes to general abilities. To see that, note, first, that there are lots of intrinsic features we vary when considering an agent's general abilities: we vary the broken leg when thinking about the ability to jump, the current nervousness when thinking about the ability to meditate, and the drugged brain when thinking about the ability to solve a mathematical puzzle. We therefore certainly don't hold *all* intrinsic features of an agent fixed when it comes to general abilities. Of course, Berofsky and Vetter do not suggest that we do. What they seem to be saying is just that we hold *only* intrinsic features of the agent fixed.

But that, too, seems very questionable. There are clear examples for what we would intuitively count as general abilities which do not depend on intrinsic properties of the agent alone. The ability to baptize. The ability to impress John. The ability to recognize a particular building. All of these abilities are fully general in the sense that we do not have a particular situation in mind when ascribing them and having them requires that the agent manage to exercise them across a large range of circumstances. Yet, all of these abilities depend as much on extrinsic properties of the agent as they depend on her intrinsic properties. The ability to baptize depends in part on the social practices in which the agent is embedded. The ability to impress John depends a great deal on John. And the ability to recognize a building depends to a large extent on the look of the building. As it seems, then, the relation between the generality of an ability and its dependence on intrinsic properties of the agent is not as tight as Vetter and Berofsky suggest.

That is why I have stated my account of general abilities in terms of stable, mostly intrinsic features of the agent. Stability, I submit, is what plays the key role. When we are thinking about what the agent can do independently of her situation, what we are holding fixed are features that the agent retains throughout a great number of situations. Those features will very often be intrinsic to the agent. But they need not be intrinsic to her; all that matters is that they are stable in the sense that the agent retains them across great changes of circumstance.

Which of the stable, mostly intrinsic features of an agent we hold fixed will vary. We may be interested in an agent's general ability sing on stage and hold fixed that the agent is so anxious of singing on stage that she immediately freezes as soon as she goes on stage. But we may also vary her nervousness. Each time, we are interested in one of the agent's general ability: an ability she has in view of some relevant set of stable, mostly intrinsic features of hers.

4.5.2 Specific abilities

Let's now look at specific abilities. We know that a specific ability is an ability to perform some action in some particular situation. I said before that this means that the facts of the situation will be held fixed in the modal base. But just which and how many of the facts of the situation have to be held fixed? Here, one can take various routes. As a result, one will get a whole spectrum of varyingly specific abilities, all of which are legitimate in their own right, and none of which deserve privilege over the others.

Out of these many varyingly specific abilities, I will pick out three, because I take them to be particularly interesting. There are what I call *particular abilities*, which are abilities in view of the *totality* of facts obtaining at some specific time. There are what I call *conative abilities*, which are abilities in view of all but a few facts obtaining at some specific time – certain mental states of the agent are left undetermined in the set of facts that go into the modal base. Finally, there are what I call *opportunities:* abilities in view of the external facts, where it will to some extent be a matter of context which facts count as external. As we'll see, there is room for more classes of specific abilities, but the three I just mentioned seem to me to be the most prevalent ones.

Suppose I intentionally jump a fence. In a very good sense of "specific ability", my intentionally jumping the fence proves that I had the specific ability to jump the fence. For this to be so, I need not have a general ability to jump that fence; my jump a moment ago may have been the first and the last time I was and will ever be able to succeed – a wondrous fluke, a marvelous stroke of sheer luck. Yet, I did it, and so I was obviously able to do it at that moment.

This interpretation of specific abilities, though quite extreme in some ways, is quite well established in the literature. It is the sense of ability according to which each agent was able to do whatever they intentionally did and unable to do what they intended but failed to do. It is the sense of ability Geach has in mind when he says that "what a man does, he can do" (Geach 1957: 15). Following Honoré, it can be dubbed a "particular ability":

> To summarize the use of 'can' (particular) in relation to particular actions: success or failure, on the assumption that an effort has been or will be made, is the factor which governs the use of the notion. If the agent tried and failed, he could not do the action: if he tried and succeeded, he was able to do it. (Honoré 1964: 464)

Particular abilities are fully in line with SUCCESS$_{AA}$. They are abilities in view of the agent's total situation: what we need to hold fixed is the totality of features of the agent's circumstances. In the case of particular abilities, the relevant possible situations will narrow down to just one: the actual situation.[9] The modal success rate will therefore become an all or nothing matter. If the agent intends to φ and succeeds, she has the particular ability, because she will thereby φ in a sufficient proportion – *all* – of the relevant intention situations. If she intends to φ

---

[9] I am assuming determinism here and throughout my discussion of particular abilities.

but fails, she lacks it, because she will not ϕ in *any* of the relevant intention situations.

What if the agent cannot intend to ϕ in the first place in the actual world? Then she will lack the particular ability. Recall that one way for SUCCESS$_{AA}$ to be violated is for the relevant possible situations not to contain any intention cases to begin with. This is just what the existential requirement states. For an agent to have an ability, there have to be relevant intention situations – it has to be possible for the agent to intend to ϕ to begin with. If the relevant possible situations narrow down to the actual situation, as they do in the case of particular abilities, an agent therefore lacks the ability, if the actual situation is not an intention situation.

At first sight, this may strike you as strange. Come to think of it, though, it becomes apparent that it is not. To see that, note that we are talking exclusively about agentive abilities so far. If a situation is such that it is impossible for the agent to intend to ϕ in it, then it seems just right that the agent also cannot perform an act of ϕ-ing in that very situation. *In view of the fact* that the agent cannot (in the properly restricted sense) intend to ϕ, the agent cannot perform an act of ϕ-ing. This is all as it should be. Actions are behavioral episodes brought about by an intention. Without an intention, no action can ensue.[10] SUCCESS$_{AA}$, including the existential requirement, yields the right verdict here.

I said that particular abilities are rather extreme in some ways. And they are. Specific abilities are abilities to ϕ in a particular situation. But of course, a situation can be individuated more or less finely. In the case of particular abilities, the situation is individuated in the finest possible way, namely by the totality of facts obtaining in a certain time span. We are holding absolutely every feature of the agent's circumstances fixed.

That need not be so, though. We may just as well think of a less extreme interpretation of specific abilities. We can be interested, for instance, in what the agent can do in the situation she is in, and yet abstract away from certain features of that situation. We may care about an agent's ability to swim here and now, say, but abstract away from the fact that the agent is strongly disinclined to swim (she does not want to get her hair wet and sees no reason in favor of swimming, say).

In the sense of a *particular* ability, the agent lacks the ability to swim in such a case. In view of her strong disinclination towards swimming, the agent cannot

---

10 Note that the notion of an intention at work here is not particularly challenging. The agent does not have to be aware of the intention, and the intention does not have to be a conscious state.

form the intention to swim. But in a slightly less extreme sense, she may well have the specific ability: abstracting away from the fact that the agent has no inclination to go swimming, it may well be that there is a sufficiently high proportion of relevant intention situations in which the agent swims. Her intentional state is not held fixed, after all. Hence, the relevant possible situations will contain quite a few intention situations. And among those, there may well be quite a few in which the agent effectively swims.

For another example, consider a person who holds a firecracker and a lighter in her hands, but is anxious to light the firecracker. Does that person have a specific ability to light it? In the sense of a particular ability, she does not – in view of her current anxiety, she cannot intend to light it. There will not be any intention situations among the relevant possible situations. But according to a slightly less extreme interpretation, she *does* have the specific ability. Abstracting away from her current anxiety to light the firecracker, she can light it in the situation she is in. Holding fixed virtually every fact apart from her current anxiety, there are relevant possible situations in which the agent intends to light the fire cracker, and in a sufficient proportion of those situations, the agent will effectively light it.

Are those still specific abilities we are thinking about? I should think so. We are, after all, asking what the agent can do here and now, in some particular situation. The mere fact that we are abstracting away from *some* of the features of the situation – most commonly, the agent's temporary mental states – does not show that we are already in the realm of general abilities altogether. Specific abilities, I should think, can vary with respect to the completeness of the facts of the actual situation that go into the modal base. We can abstract from a small fraction of the features of the actual situation and yet be talking about an ability the agent has to ϕ in the very situation she is in.

I would even go as far as saying that we *usually* abstract from the facts determining the agent's motivational states when thinking about an agent's specific abilities. In this, I side with Berofsky, who writes that

> [a] person who has the power to act may fail to do so for reasons having to do with her motivation or her will. (...) We normally distinguish between these [conative conditions] and the other conditions (ability, opportunity), regarding only the latter's absence as depriving one of the power to act. (Berofsky 2002: 196)

What Berofsky points out in this passage is that we normally do not count an agent as unable to perform some action in a particular situation just because the person is temporarily disinclined to perform that action or has other action-preventing motivational states. I take this to be exactly right. Specific abilities whose modal base does not contain the motivational states of the agent are therefore an important class of specific abilities. Following Berofsky's choice of words, I will call specific abilities of this kind "conative abilities".

To see how common conative abilities are, note that it is abilities of that kind which seem to underlie much of our thinking about whether or not agents could have done otherwise. "Could have done otherwise"-locutions create a context in which it is the agent's specific abilities, and not her general abilities, which are at issue. When I say "I could have asked for help", I am not just saying that I have the general ability to ask for help. Rather, I state that I had the ability to ask for help in the very situation I am picking out by my use of the past tense. It is specific abilities that matter. But the way we actually talk about things people could have done differently suggests that it is not particular abilities that are at issue in such contexts.

Take the disinclined swimmer once more. Usually, we will want to say that the agent could have swum – she just wasn't motivated to. Likewise in many other cases in which the agent's motivation is what prevents the act: the vegetarian could have eaten the meat, she just wasn't motivated to. The moral person could have stolen the wallet, she just wasn't motivated to. I could have asked for help, I just wasn't motivated to. What this shows is that we usually abstract away from agent's motivations when thinking about what the agent could have done differently. Since "could have done otherwise" locutions are very common in ordinary discourse, this shows that it is very often not particular, but rather conative abilities that concern us when we are thinking about agents' specific abilities.

Finally, let me talk about opportunities. What is an opportunity? Frankly, I don't think anyone has a clear idea. But I think Kenny is right when he points out that "an opportunity is something external" (Kenny 1976: 218). Thus, an opportunity for ϕ-ing, in the most natural understanding of the term, will simply be favorable *external* circumstances for ϕ-ing. The world outside the agent has to provide what it takes to perform ϕ, so to say. If that is true, then for an agent to have an opportunity will simply be for the agent to ϕ in a sufficient proportion of intention situations in which the external circumstances of the situation are held fixed, according to the success view.

But what counts as external to the agent? Does a broken leg count as a lack of opportunity? Or does it simply prevent the agent from taking an opportunity? I take this to be rather unclear. Moreover, I take it to be unclear how many of the external features of the situation have to be taken into account. Does an agent have an opportunity to hole a putt when there is a bump in the grass which the agent does not see, and which will ultimately take the ball off course? Maybe. Maybe not.

What this suggests is that the notion of an opportunity is context sensitive along two dimensions. Whether or not we count someone as having an opportunity will depend, first, on how completely the external facts of the situation go into the modal base, and secondly, it depends on which features we count as external to the agent in the first place. Depending on one's choices along those two dimensions, an opportunity may in some contexts simply come out identical to an agent's conative abilities – we hold everything fixed except the agent's motivational states. In other contexts, however, the modal base will not comprise the agent's broken leg. And this is surely something we want to hold fixed in the case of conative abilities.

Can having an opportunity also just amount to the having of a particular ability? I do not think this would be a natural interpretation of "opportunity". It would be rather odd to treat an agent's own motivational states – the agent's disinclination to swim, for instance – as something that deprives the agent of the opportunity to swim. It should be coherent, in any context, to say: "I had the opportunity, but I did not feel any inclination to take it". If the agent's motivational states entered the modal base of an opportunity in some contexts, then this would not be truly utterable in that context. I take it to be settled, therefore, that "opportunity" carries an at least somewhat external connotation and that the minimal requirement for circumstances to be external to an agent is for the circumstances not to comprise the agent's own motivational states.

4.5.3 Dependence

So far, we have looked at general and specific abilities individually. Let's now check how far the two depend on one another. We know that an agent can, in a given situation, have a general ability to φ without having the specific ability to φ in that very situation. This is one of the essential insights the distinction between the two is built upon. But what about the other way around? Does the having of a specific ability require that the agent also has the corresponding general ability?

Some authors seem to think so. Berofsky's concept of specific abilities (or token-abilities, as he calls them), for instance, seems to be simply that of an opportunity to exercise one of one's general abilities (Berofsky 2002: 196). To have a specific ability, then, the agent would have to have the general ability first. Whittle objects to this by pointing out that an agent may be perfectly able to jump 5 feet high in very specific circumstances, while at the same time being unable to jump 5 feet high across a large range of circumstances (Whittle 2010). This suggests that an agent may very well have a specific ability to ϕ without also having the corresponding general ability as well.

Who is right? Again, that seems to me to depend on one's use of "specific ability". The three types of specific abilities I characterized in the last section do not seem to require that the agent have a corresponding general ability.

Take particular abilities. My intentionally jumping the fence proves that I had the particular ability to jump the fence, irrespective of my having or lacking a general ability to jump the fence. My jump a moment ago may have been the first and the last time I was and will ever be able to succeed. Yet, I did it, and so I was obviously able to do it at that moment.

Likewise, in the case of conative abilities. Suppose a child holds a firecracker in her hand but has no intention to light it. Suppose further that the lighter she holds in her hand is extremely easy to handle, whereas the child is unable to operate most lighters. In that case, she has the conative ability, because she lights the fire cracker in a sufficiently high proportion of the relevant possible situations in which she intends to light it. The relevant possible situations will only contain situations that are like the actual situation in the non-motivational aspects of the situation, among them the fact that the child holds the easy-to-handle lighter in her hand.

Yet, the child lacks the general ability to light a firecracker, because in the vast majority of intention situations relevant for *that* ability she fails to light a firecracker. That is because the relevant possible situations will this time contain any situation in which the child is as in actuality with respect to her stable, intrinsic features. In the vast majority of those situations, she will get hold of lighters that are harder to operate. And hence, she will not light the firecracker in most of them.

Again, likewise in the case of opportunities. There is very good sense to be made of a once in a lifetime opportunity for committing the perfect crime – the circumstances being favorable this one time for getting away with what one did. Yet, this does not entail that the agent has a general ability to commit the perfect crime. Only some specific abilities will depend on the having of a general ability.

That is not to say that there is no good sense of "specific ability" which requires a corresponding general ability. We can define a *solid particular ability* as a compound of a general and a particular ability. Likewise, we can define a *solid conative ability* as a compound of a general and a conative ability. Finally, we can define a *solid opportunity* as a compound of a general ability and an opportunity, thus capturing what Austin (1956: 218) famously called the "all-in" sense of ability. As far as I can see, these are perfectly legitimate senses of "specific ability". And those senses do of course depend on the agent having the relevant general ability at the outset.

What about the other way around? When I first introduced the distinction between general and specific abilities (→ 1.2), I said that it is taken for granted that one can have a general ability without having a corresponding specific ability. In *that* sense, general abilities do not depend on specific abilities. But at the same time, it does not seem very plausible that one should have a general ability to ϕ without there being *any* possible situation in which one also has the specific ability to ϕ. Quite to the contrary! When I have the general ability to sing, say, this seems to entail that there are various possible situations for which it is true that I have the specific ability to sing in those very situations. In this sense, general abilities *do* depend on the specific abilities.

This way of thinking about the relation of general and specific abilities is implicit in a number of passages in the literature, and in fact it seems to underlie the whole idea of the reliability of general abilities. When Whittle says that a specific ability requires that the agent be able to do something in particular circumstances, whereas a general ability requires that she be able to do it in a large range of circumstances (Whittle 2010), what she is saying is basically that one general ability goes along with many specific abilities. Maier is even more explicit on his commitment to such a picture: on his view, a general ability is actually constituted by a certain number of specific abilities – abilities to do something in particular situations (Maier 2013).

Maier and Whittle are on the right track here. General abilities to ϕ do require that there are situations in which the agent has a specific ability to ϕ. I also think that it makes sense to think of general abilities as being made up of specific abilities.[11] It would be overly complicated to go through all of the interpretations of specific abilities to show to what extent the having of general abilities depends on the having of each of the different kinds, but it can easily be shown that Whittle's and Maier's remarks make sense when it comes to particular abilities.

The argument for the claim that to have a general ability to ϕ, there have to be some situations in which the agent has the particular ability to ϕ, is very straightforward. Take the general ability to light a match. To have that ability, there will obviously have to be quite a few situations in which the agent has the particular ability to light a match – in which she intends to light one and succeeds. If there were no such situations, then SUCCESS could not be met for the general ability to light a match to begin with. It is only met, after all, if the pro-

---

11 Unless, of course, specific abilities are interpreted solidly, i.e. such that they entail the having of a general ability. When I speak of specific abilities, I mean the specific abilities of the non-solid kind, unless otherwise noted.

portion of the relevant intention situations in which the agent lights a match is high enough. If there are *no* intention situations in which the agent intends to light the match and lights it, then this cannot be true. A general ability to φ therefore entails that the agent also has a particular ability to φ in at least some situations.

Let me emphasize that those situations – the situations in which the agent has the particular ability to φ – need not be actual. It is perfectly possible for an agent to have a general ability to light a match without having a particular ability to light a match in any actual situation. One need not ever actually intend to φ and φ in order to have a general ability to φ, after all. Perhaps there is always reasons for me not to intend to light a match. Or there is never a match around. All this is perfectly compatible with my having the general ability to light a match.

To what extent can general abilities be thought of as being constituted by particular abilities? Again, the line of reasoning is quite straightforward: we can think of a an ability to φ as being constituted by the set of the relevant possible situations in which the agent intends to φ and φ's. A particular ability is then constituted by exactly one set of situations – the one relevant possible situation in the modal base in which the agent intends to φ and does φ. General abilities are correspondingly constituted by the sets of situations in their modal bases in which the agent intends to φ and φ's. Each such situation constitutes a particular ability to φ in those very circumstances. Hence, general abilities can be seen as being constituted by particular abilities.

4.5.4 Success and failure

A very controversial issue in the literature is how specific abilities relate to individual failure and success. In the case of *general* abilities, it is obvious that such abilities can be retained, even when an actual intention to exercise them is not successfully realized. This can be put in terms of the *maskability* of general abilities. General abilities can be masked: my ability to swim can be masked by a cramp in my foot, my ability to sing can be masked by my temporary hoarseness, my ability to hit the bull's eye can be masked by me being distracted, and so forth. All of those things prevent my intention from being successfully realized, but they do not deprive me of my general ability. General abilities are not lost in virtue of a singular failure, then.

Reversely, an individual success does not *bestow* an agent with a general ability. The agent may successfully φ in a specific situation and yet she may not have a general ability to φ. This has to do with the reliability of general abilities. To have such an ability, the agent has to be able to perform successfully in a variety of situations. A singular success cannot guarantee any such thing.

The interesting question is how *specific* abilities relate to failure and success. There are two questions here. First: does an individual success show that the agent had the specific ability to ɸ in the situation she was in? And secondly: is the having of a specific ability compatible with failure? Both questions are highly controversial within scholarly discourse.

As to the first question, many think that if an agent ɸ's, then this suffices to show that the agent had the specific ability to ɸ. Austin, for instance, holds that when a golfer holes a putt, "it follows merely from the premise that he does it, that he has the ability to do it, according to ordinary English" (Austin 1956: 218). Mele (2002) thinks that things are more complicated than that. And Whittle (2010) thinks that there is a sense of "specific ability", which requires that the agent be able to reliably replicate her action under the very same circumstances. I am not quite sure what it would take for a subject to be able to replicate an act under the exact same circumstances, but Whittle clearly commits to the view that a singular success does not suffice for the having of certain kinds of specific abilities. Things are controversial, to say the least.

Likewise for the second question. Does a failure to ɸ show that the agent did not have the specific ability to ɸ? Or are specific abilities maskable in the sense that they can be retained even if the agent tries and fails? Here, too, intuitions vary. Fara (2008) is clear on the matter: he thinks specific abilities can be masked. Whittle (Whittle 2010) thinks that some fairly specific abilities can be masked, but that the most specific abilities cannot. A third stance would be the view that if you try and fail, this shows that you lacked the specific ability to ɸ (von Wright 1963: 61).

We can note that the connection between specific abilities and success and failure is far from clear. There are quite a few ideas floating around, and thus far, there is little understanding of the way they relate and which of the conflicting ideas captures the primary notion of specific abilities, if there is such a thing. Against the background of the various distinctions we drew between various kinds of specific abilities, however, it should by now be clear that a lot hinges on one's understanding of a specific ability here.

Particular abilities are maximally closely related to failure and success. An agent has a particular ability in a certain situation if and only if she intends to ɸ and succeeds. If she intends to ɸ and fails, she lacks the particular ability. An individual success therefore bestows, and an individual failure deprives the agent of a particular ability. This entails that particular abilities are not maskable. Whatever it is that prevents the agent from successfully ɸ-ing goes into the modal base. It therefore deprives her of the ability to begin with and does not, as in the case of general abilities, mask it.

Things are more complicated in the case of conative abilities. Here, we vary whether or not the agent intends to φ. Hence, the relevant possible situations will contain some in which the agent intends to φ and some in which the agent does not intend to φ. The agent has the conative ability if and only if she φ's in a sufficient proportion of the relevant intention situations. Whether or not an actual success suffices for the agent to have an ability and whether or not an actual failure suffices for the lack of the ability depends crucially on the standards for what counts as a sufficient proportion. Perhaps the agent has to φ in *all* of the relevant intention situations in order to count as having a conative ability in a context. If this is the case, then an actual failure deprives the agent of the ability, but an actual success does not bestow her with it. If the agent has to φ in one of the relevant intention situations, in contrast, then it is the other way around. If she has to φ in a few, neither is the case. So here, context will play a crucial role. The same, I take it, goes for opportunities.

4.5.5 Degrees of specificity

General and specific abilities differ in the kinds of facts that go into their respective modal bases. In the case of general abilities, we hold only the stable, mostly intrinsic features of the agent fixed. In the case of specific abilities, we also hold various features of the situation fixed. By now we know that the kinds of facts that go into the modal base also vary *within* the class of general abilities and *within* the class of specific abilities.

In the case of the general ability to shoot a rabbit, we may hold only the agent's properties underlying the agent's shooting skills fixed or we may additionally hold her moral reservations against shooting animals fixed. In the case of the general ability to juggle, we may want to hold fixed that the agent is currently in a coma, or we may vary this fact and only hold the properties underlying the agent's juggling skills fixed. And so on. Likewise for specific abilities. We may hold the totality of the facts about the agent and her situation fixed, we may vary the agent's motivational states, we may vary certain facts about the external situation, or vary both some motivational states and some external facts.

Intuitively, this seems to indicate that abilities form some sort of scale, along which various abilities to φ can be ordered in accordance with their specificity and generality, respectively. The more facts are held fixed, the more specific is an ability. The less facts are held fixed, the more general. Or so it seems.

Actually, things are a bit more complicated. As I will argue, there is indeed sense to be made of an ordering of abilities according to their degree of specificity, but first of all, that ordering will only be partial (not any two abilities to φ stand in the "is more specific than" relation to one another), and secondly, the sheer number of facts that are held fixed across situations is not the proper criterion for the ordering. Let me take these points in reverse order.

Seeing that the number of facts is not the crucial criterion for specificity is easy. It is, for instance, not at all clear why my ability to do a handstand with a broken arm (one fact?) should be less specific than my ability to do a handstand on a stage while being drunk (two facts?). The sets of facts that are held constant in both cases are simply different. It is not reasonable to think that their sheer number has anything to do with the ability's specificity.

Yet, some abilities *do* seem to be more specific than others. My ability to sing on stage when Adele sits in the audience seems to be more specific than my ability to sing on stage, for instance. What seems to matter, then, is that the facts that are held constant are properly related. In which way? One idea would be to say that

> SPECIFICITY$_{PRELIMINARY}$. an ability A1 is more specific than another ability A2, if and only if the set of facts held fixed in the case of A2 is a proper subset of the set of facts held fixed in the case of A1. If A1 is more specific than A2, then A2 is more general than A1.

This yields the right results in the cases we looked at. My ability to do a handstand with a broken arm does *not* come out more specific than my ability to do a handstand on stage while being drunk, whereas my ability to sing on stage when Adele is present *does* come out more specific than my ability to sing on stage.

But we need to take one further subtlety into account. I take it that we want to be able to say that *your* ability to sing on stage when Adele is present is more specific than *my* ability to sing on stage. And this does not follow, if the facts we are holding constant in my case have to be a proper subsets of the facts we are holding fixed in your case. We will, among other things, hold your larynx system fixed in your case while holding my larynx system fixed in my case, after all.

This suggests that it is not actually the facts themselves which have to be related in the right way in order to yield degrees of specificity, but *kinds* of facts. What we are holding fixed are facts about the *agent's* larynx system. That means we are holding facts about you fixed in your case and facts about me in my case. This suggests that

> SPECIFICITY. an ability A1 is more specific than another ability A2 if and only if the kinds of facts that are held fixed in the case of A2 are a proper subset of the kinds of facts that are held fixed in the case of A1. If A1 is more specific than A2 then A2 is more general than A1.

It should by now be clear why we will not end up with a full linear ordering of abilities to φ. It is simply not the case that any two abilities to φ will stand in the "is more specific than" relation to one another, if specificity is spelled out in the way I just proposed. The ability to do a handstand with a broken arm and the ability to do a handstand on stage while being drunk, for instance, are clear examples of abilities that do not match up neatly with respect to their respective degrees of specificity.

4.5.6 Hypothetical circumstances

There is one rather intricate issue we have to turn to before closing the section on general and specific abilities: we need to turn to the role of hypothetical circumstances. To see what I am getting at, take the ability to do a handstand with a broken arm as opposed to the ability to do a handstand. On the account of general and specific abilities I have just given, the former is more specific than the latter. In the case of the former, we hold the agent's broken arm fixed; in the case of the fully general ability, we don't.

Now let's suppose the agent does not actually *have* a broken arm. Then we obviously cannot think of the broken arm as something we hold fixed. What we hold fixed, across possible situations, are always features of the actual situation. Hence, the distinction between the ability to do a handstand and the ability to do a handstand with a broken arm cannot, in such a case, be captured in terms of a difference in the facts we hold fixed, with the broken arm being among the fixed facts in one case but not in the other. And that means that the account of degrees of specificity cannot be applied to the distinction between the ability to do a handstand and the ability to do a handstand with a broken arm.

This points to a more general issue. Earlier, we saw that, according to Kratzer, the modal base of an ability is always realistic. We hold certain facts of the actual situation fixed across the possible situations, and hence the actual world will always be contained in the situations which make up the modal base. This is just as it should be: as Kratzer so insightfully taught us, agents have abilities *in view of* certain facts. The modal base is the set of situations in which those facts are held constant – it is the set of situations in which the facts obtain in view of which an agent has or lacks an ability.

What the example of the broken arm shows, however, is that *hypothetical* circumstances – circumstances that are not facts, but non-actual states of affairs – play a vital role in our thinking about abilities as well. We may be interested in an agent's ability to do a handstand with a broken arm, no matter whether or not the agent actually has a broken arm. We may be interested in an agent's ability to run a marathon in hot weather, no matter whether the runner is actually located in Sumatra or at the North Pole. And we may be interested in an agent's ability to jump on one foot when completely drunk, even if the agent has never had a sip of alcohol. Hypothetical circumstances enter into our thinking about abilities quite often, then. The question I will pursue in this section is: how do they enter?

The answer is a bit complicated, because hypothetical circumstances can enter in a variety of ways. So far, we have thought about the role of circumstances only in terms of the facts about the agent and her situation that are held fixed across the possible situations – the sets of circumstances in view of which the agent has the ability. That is to say that the structure of ability statements, as we have thought of it so far, looked like this:

CAN. S can, in view of C*, ɸ.

C* is the set of circumstances in view of which the agent has the ability. That set of circumstances is always factual: only facts – states of affairs which obtain at the actual world – can go into C*. Since C* determines the modal base of an ability, the modal base of abilities is always realistic: the actual situation is always among the modal base situations.

In fact, however, the structure of abilities is way more complex. As I will lay out in what follows, ability statements have four, instead of just one, placeholders for circumstances. Really, they have the following structure:

CAN*. In C1, S can, in view of C* and when C3, ɸ-in-C2.

As before, C* is the set of circumstances in view of which the agent has the ability. This set of circumstances is factual. It determines a realistic modal base. C1, C2, and C3 are further sets of circumstances. Importantly, all of those circumstances can be purely hypothetical. To see what I am getting at, let's look at an example of an ability ascription which features hypothetical circumstances and see how it can be interpreted. Let's assume that Fred is actually in a place with cold weather. On this assumption, let's see how we can interpret a statement like the following:

(A) Fred can, in view of his physical constitution (C*), run a marathon *in hot weather.*

As I will argue, there are three distinct interpretations available, depending on whether the hypothetical circumstances – Fred being in hot weather – enter the ability ascription via C1, C2 or C3. Here is the first interpretation:

(INT-1) In hot weather (C1), Fred has, in view of his physical constitution (C*), the ability to run a marathon.

In INT-1, the hypothetical circumstances are outside the scope of the ability ascription. What is stated is that Fred has a certain ability under certain circumstances. A statement of this form does not imply that the agent *actually* has the ability to run a marathon. It is a statement about an ability that is had under certain conditions, and it entails a statement about the agent's abilities in actuality only if the conditions specified in C1 are actual and not merely hypothetical.

The statement can therefore also be interpreted in the subjunctive mode. It can be read as the statement that *if the weather were hot*, Fred *would* (in view of his physical constitution) have the ability to run a marathon. If we translate this into the framework of the success view, INT-1 states the following:

(INT-1$_{SUCCESS}$) If Fred were in hot weather (C1), then Fred would run a marathon in a sufficient proportion of the possible situations in which Fred has his actual physical constitution (C*) and intends to run a marathon.

The hypothetical circumstances give a condition for the having of an ability, which is itself structured as before. Within the scope of the ability operator, hypothetical circumstances play no role.

For the second interpretation, let's assume that the hypothetical circumstances enter the ability ascription via C2. That is to say: they enter the ability ascription in the form of a specification of the action type the agent is said to be able to perform – in a statement of the form "S can ϕ", they qualify "ϕ". If we interpret the statement along those lines, it states the following:

(INT-2) Fred has, in view of his physical constitution (C*), the ability to run-a-marathon-in-hot-weather (C2).

The hyphens indicate an action type: running-a-marathon-in-hot-weather is a different action type from just running a marathon. The hypothetical circumstances do the same job as any other action type qualifier: running-in-hot-weather functions like running slowly, running on one leg, or running in circles. In contrast to the first interpretation, the second interpretation *does* entail that the agent actually has the ability in question. It states that Fred has, in view of his physical constitution, the ability to perform a certain action type, namely running-a-marathon-in-hot-weather.

How is INT-2 accommodated by the success view? According to the success view, S has an ability to φ if and only if the agent φ's in a sufficiently high proportion of relevant possible situations in which the agent intends to φ. Since φ-ing corresponds to running-a-marathon-in-hot-weather in INT-2, INT-2 therefore states the following:

> (INT-2$_{SUCCESS}$) Fred runs a marathon-in-hot-weather (C2) in a sufficiently high proportion of the possible situations in which Fred agent has his actual physical constitution (C*) and intends to run-a-marathon-in-hot-weather (C2).

Running-a-marathon-in-hot-weather is treated just like any other qualified action type, then. An agent has the ability to run on one leg, according to the success view if and only if the agent runs on one leg in a sufficient proportion of relevant possible situations in which the agent intends to run on one leg. Running-a-marathon-in-hot-weather is accommodated in the very same way.

Finally, let's turn to the third interpretation: here, we want to express that Fred has, in view of his physical constitution, the ability to perform the action type of running a marathon when certain hypothetical circumstances obtain – namely, when the weather is hot. We are stating the following, then:

> (INT-3) Fred has, in view of his physical constitution (C*) and when the weather is hot (C3), the ability to run a marathon.

That this makes sense and is an important way of interpreting the ability statement in question (and ability statements containing hypothetical circumstances quite generally) becomes most apparent when we think of contexts in which two agents' abilities to perform one and the same action type are compared. Suppose I say, bracketed terms being implicit: "Frieda can (in view of her physical constitution) run a marathon". And you say "Yeah, but Fred can (in view of *his* physical constitution) run a marathon even in hot weather". Then it seems to me that, intuitively, the best way to make sense of this conversation is in terms of the third interpretation I have suggested.

Your statement is definitely not intended to state, along the lines of the first interpretation, that there are certain hypothetical circumstances in which Fred has one and the same ability that Frieda has in actuality. You seem to be committing to Fred having an ability in actuality as well, namely the ability to run a marathon under certain hypothetical circumstances.

Likewise, I believe, it is also not very useful to think of your statement as saying that Fred has the ability to perform a rather different action type than Frieda. Rather, what you seem to be saying is that Fred can perform one and the same action type as Frieda, but he can do so under more conditions – while Fred can run a marathon in situations in which the weather is hot, Frieda cannot.

How can the third interpretation be accommodated by the success view? My suggestion is that INT-3 should be understood along the following lines:

> (INT-3$_{\text{SUCCESS}}$) Fred runs a marathon in a sufficiently high proportion of the possible situations in which (i) his physical constitution is as in actuality (C*), (ii) he is in hot weather (C3), and (iii) he intends to run a marathon.

On this interpretation, the hypothetical circumstances impose a further restriction on the situations among which we check for intention situations, among which, in turn, we check for performance cases. We first restrict the possible situations to the ones in which the relevant facts obtain, yielding a realistic modal base with the actual situation in it. But then we impose a second restriction. We restrict the situations further to the ones in which certain hypothetical circumstances obtain. In this step, the actual situation gets kicked out.

INT-3 yields a modification of some of the things I have about the set of relevant possible situations. So far, I have identified the relevant possible situations with the situations in which the facts obtain in view of which an agent has an ability. When hypothetical circumstances enter an ability statement via C3, however, the relevant possible situations will in fact be a subset of those situations: those in which those hypothetical circumstances obtain as well.

Quite often, hypothetical circumstances do not play a role. We may, after all, simply be interested in an agent's ability to do things in the very circumstances she is actually in. We may be interested in an agent's ability to run a marathon in hot weather when the agent is actually in hot weather. In such cases, our thinking about what an agent can do is simply a thinking about what the agent can do in view of certain intrinsic or extrinsic features of hers – being in hot weather, say.

But, and this is where hypothetical circumstances become relevant, we may also be interested in an agent's ability to run a marathon in hot weather when the agent is actually in a cold environment. In that case, the restriction to hot weather situations has nothing to do with the features in view of which the agent has the ability. The hypothetical circumstances come into play from the outside, so to speak. The relevant possible situations are a subset of the relevant possible situations in which facts of the actual situation are held fixed in that case.

What I contend, then, is that ability statements will sometimes have to be read as statements of the form "In view of C* and when C3, S can φ", where C3 specifies a contextually determined set of circumstances, which may well be hypothetical. When we ask whether some agent has the ability to run a marathon in hot weather, we will always have to restrict the realm of possible situations to the ones in which the weather is hot. But for the agent in Sumatra, that comes down to a restriction to situations that are similar to hers with respect to the weather conditions, whereas for the agent at the North Pole it amounts to a restriction to situations that differ from hers with respect to that feature.

Let us take stock. What I have laid out in the last paragraphs are three ways in which hypothetical circumstances can enter ability ascriptions. Depending on how they enter, we get different interpretations of an ability statement with a reference to hypothetical circumstances: that a certain ability is had under certain hypothetical circumstances (INT-1); that the agent has an ability to perform some-action-in-certain-hypothetical-circumstances (INT-2); or that the agent has an ability to perform some action when certain hypothetical circumstances obtain (INT-3). Each of those interpretations is accommodated differently by the success view.

What does all this have to do with general and specific abilities? I have said, in the last section, that

> SPECIFICITY. an ability A1 is more *specific* than another ability A2, if and only if the kinds of facts that are held fixed in the case of A2 are a proper subset of the kinds of facts held fixed in the case of A1. If A1 is more specific than A2, then A2 is more *general* than A1.

Since specificity is formulated in terms of facts that are held fixed, the definition does not accommodate cases in which hypothetical circumstances play a role. This may strike you as mislead. The ability to run a marathon in hot weather, you may think, is obviously more specific than the ability to run a marathon, no matter whether or not the agent is actually in hot weather. The suggestion would then be that specificity can come about via any of the sets of circumstances which can figure in an ability ascription. Specificity via the actual facts we hold fixed would turn out to be but one way among many in which specificity can come about, according to that account.

Basically, I think this is exactly right. Yet, I will reserve the terms "specific" and "general" for ways in which different sets of *actual* circumstances can relate within ability ascription, as spelled out in SPECIFICITY. In the remainder of this section, I will give some definitions for the various similar relations – relations between further sets of circumstances, C1, C2, and C3, that can figure in ability ascriptions. For reasons of taxonomical clarity, each of these relations will be dubbed differently. Besides specificity, abilities can differ in their dependency, complexity, completeness, determinacy, and locality.

First, dependency. Let us say that

> DEPENDENCY. an ability A1 is more *dependent* than an ability A2 if and only if the kinds of circumstances under which A2 is had are a proper subset of the kinds of circumstances under which A1 is had. If A1 is more dependent than A2, then A2 is more *independent* than A1.

Dependency comes about when varyingly inclusive sets of circumstances enter an ability ascription via C1. The more inclusive the set of circumstances on which the having of an ability depends, the higher the degree of dependency. DEPENDENCY thus yields that "In hot weather and with a broken leg, S has the ability to run a marathon" ascribes a more dependent ability than "In hot weather, S has the ability to run a marathon".

Further, let us say that

> COMPLEXITY. an ability A1 is more *complex* than an ability A2 if and only if the kinds of circumstances that go into the action description in the case of A2 are a proper subset of the kinds of circumstances that go into the action description in the case of A1. If A1 is more complex than A2, then A2 is more *simple* than A1.

Complexity comes about when varyingly inclusive sets of circumstances enter an ability ascription via C2. The more inclusive the set of circumstances that determine the action type of the ability ascription, the more complex the ability. COMPLEXITY thus yields that "S has the ability to run a marathon-in-hot-weather-and-with-a-broken-leg" ascribes a more complex ability than "S has the ability to run-a-marathon-in-hot-weather".

Further, let us say that

> DETERMINACY. an ability A1 is more *determinate* than an ability A2 if and only if the kinds of hypothetical circumstances that (partly) determine the relevant possible situations in the case of A2 are a proper subset of the kinds of hypothetical circumstances that (partly) determine the relevant possible situations in the case of A1. If A1 is more determinate than A2, then A2 is more *indeterminate* than A1.

Determinacy comes about when varyingly inclusive sets of circumstances enter an ability ascription via C3. The more inclusive the set of kinds of hypothetical circumstances that partly determine the relevant possible situations, the higher the degree of completeness. DETERMINACY thus yields that "S has the ability to run a marathon when the weather is hot and S has a broken leg" ascribes a more determinate ability than "S has the ability to run a marathon when the weather is hot". Since the circumstances that enter via C3 combine with the facts that are held fixed in an ability ascription (C*) to determine the relevant possible situations, a higher degree of determinacy will, *ceteris paribus*, yield a more exclusive set of relevant possible situations.

Finally, the relevant possible situations as a whole may be more exclusive in the case of one ability than in the case of another. I will speak of locality in that case. Let us say that

> LOCALITY. an ability A1 is more *local* than an ability A2 if and only if the circumstances, actual and hypothetical, that determine the relevant possible situations in the case of A2 are a proper subset of the circumstances, actual and hypothetical, that determine the relevant possible situations in the case of A1. If A1 is more local than A2, then A2 is more *global* than A1.

4.6 Masks

So far, we have seen that the success view delivers an account of degrees of abilities as well as of general and specific abilities and that it circumvents the problem of the impeded intention. In this section, I will argue that it also circumvents the problem of masks, as it emerged in connection with the simple conditional analysis.

According to the simple conditional analysis, an agent has an ability to φ if and only if the agent would φ, if she intended to. For the simple conditional analysis, the problem with masks is that there are cases in which an agent has the ability to perform some action, and yet it is not true that the agent would perform the action if she intended to perform it. In fact, it can pretty much always be the case that the world is such that something would interfere with the exercise of the intended performance. A sudden gust of wind can blow the golf ball off course and prevent the able agent from holing the putt. A broken leg can prevent the swimming champion from actually swimming. And a cramp in a gymnast's arm can prevent her from doing the handstand she has so often done before.

The problem has to do with the truth conditions of the counterfactual. A counterfactual is true if and only if the closest worlds in which the agent intends are worlds in which she succeeds.[12] That means that we must depart from the actual world no further than necessary to reach an intention world. That is a problem. If the actual world happens to contain an interfering factor for the exercise of an intention on the agent's part, then this interfering factor will in all likelihood also be present in the closest worlds in which the agent intends to ϕ. Hence, the agent's intention will not be realized in that world.

Having said this, it should already have become apparent why the same problem does not arise for the success view. The success view does not just take situations into account which are, apart from the fact that the agent intends to ϕ in them, as similar to the actual situation as possible. Rather, most facts of the actual situation are completely irrelevant to the determination of the set of situations that matters to the agent's ability. When we are thinking about the golfer's ability to hole a putt, we are usually not interested in the ability to hole a putt in view of the actual wind regime. Hence, the wind regime is not held constant across possible situations.

What *is* held constant? That depends. The agent's bodily and mental constitution could be one set of features. If we then ask whether the agent holes the putt in a sufficient proportion of situations in which *those* circumstances obtain and she intends to hole it, the answer may well be "yes". This is so despite the fact that the actual situation is one in which the agent's intention would not have been successfully realized.

Masks are not a problem for the success view, then, because the mask – the interfering factor in the actual situation – is usually varied across the relevant possible situations. That means that the very few situations in which the mask is present do not in any sense have priority over situations in which it is not present. Hence, the fact that the actual situation is a mask situation is not of particular interest. The actual situation does not have any special status among the realm of the relevant possible situations.

It is worth noting that a feature that is a mask for one ability may well be a circumstance that *deprives* the agent of another, more specific ability (no matter whether the higher specificity is achieved in terms of locality or more determinateness). The wind regime, for instance, is a mask for the golfer's general ability to hole a putt, but it really *deprives* her of her ability to hole a putt in view of the actual wind regime. The agent has the former ability but lacks the latter. The reason for her lack of the latter is that we have hold a feature fixed that is merely a mask for the former ability and therefore gets varied in that case.

---

[12] Again, I neglect problems with the limit assumption.

For another example, compare the ability to do a handstand and the ability to do a handstand with a broken arm. In the case of the ability to do a handstand, broken arms usually get varied. In the case of the ability to do a handstand with a broken arm, in contrast, they are held constant. Again, the agent has the former ability, but lacks the latter. The reason for that is that the feature that figures as a mask for the former ability gets fixed in the case of the latter ability. The agent does a handstand in a sufficient proportion of the possible situations in which she intends to do a handstand and her muscular constitution, say, is as in actuality. But she does not do a handstand in a sufficient proportion of the possible situations in which she intends to do a handstand and her arm is broken.

4.7 Formal considerations revisited

This chapter is about to draw to a close. What remains to be discussed before we move on is the impact of the last objection we looked at in chapter 3, in our discussion of possibilism. As Kenny has convincingly shown, possibilism is bound to fail in virtue of the fact that the ability operator – the "can" of ability – does not obey even the most basic laws governing possibility. This section explores whether or not the same or a related objection arises for the success view as well.

Kenny's argument against possibilism, recall, goes like this. The focus is on two modal axioms, T and K2, the first of which holds for most, and the second for any possibility operator:

T. $p \to \Diamond(p)$

K2. $\Diamond(p \lor q) \to (\Diamond(p) \lor \Diamond(q))$

As Kenny shows, both T and K2 fail to state laws when $\Diamond$ is interpreted as the ability operator. From this he concludes that the "can" of ability cannot be interpreted in terms of restricted possibility.

I take it to be obvious that the argument, as it stands, does not affect the success view. The success view does not interpret abilities in terms of restricted possibility. Hence, it is not to be expected that the "can" of ability obeys the same laws as the possibility operator. Hence, to make the reverse point, a failure to meet the basic laws of possibility does not pose a problem for the success view.

That $SUCCESS_{AA}$ analyzes CAN, the "can" of ability, as something other than restricted possibility should be obvious, but let me make the differences explicit once more. According to the success view,

SUCCESS$_{AA}$. an agent S has an ability to φ if and only if S φ's in a sufficient proportion of relevant possible situations in which S intends to φ.

Compare this condition to the truth conditions for possibility statements:

◊. It is possible that p if and only if there is at least one p world among the relevant worlds.

Apart from the fact that SUCCESS$_{AA}$ is formulated in terms of situations instead of complete worlds, there are two important differences between those two conditions. First, SUCCESS$_{AA}$ imposes a second restriction on the realm of the possible situations. The possible situations get restricted not only to the relevant ones, but rather to the relevant ones in which the agent intends to φ. Secondly, and more importantly, while ◊ requires only that there be at least *one* p-world among the restricted set of worlds, SUCCESS requires that there be a sufficient *proportion* of performance cases among the restricted set of situations.

These differences are crucial – ability ascriptions behave very differently from possibility ascriptions. For that reason, it is not to be expected that they obey the modal laws governing possibility – no matter how basic those laws are in that realm.

Interestingly, Kenny seems to think that his argument shows that *any* possible worlds analysis of ability is bound to fail (1976: 226). But this strikes me as a clear *non sequitur* from the insight that abilities cannot be understood as restricted possibilities. For the inference to go through, the only available possible worlds analysis of abilities would have to be one in terms of possibility. Clearly, this assumption is misled, if the success view is on the right track. Kenny's objection from formal inadequacy does not affect the success view.

But wait a minute, you may say. So possibilism fails because it analyzes abilities as restricted possibilities and is therefore committed to the validity of certain modal inferences, which fail to hold for ability statements. And the reason this does not affect the success view is that the success view does not analyze abilities in terms of restricted possibilities and is *therefore* not committed to the validity of the same inferences. But isn't this, too, a *non sequitur*?

Agreed. Even though the success view does not explicitly analyze abilities in terms of restricted possibilities, SUCCESS$_{AA}$ may still turn out to entail the very same inference rules Kenny has shown to be invalid for ability statements. In that case, Kenny's argument would be easily adaptable as a refutation of the success view. For then we could easily argue:

i. If SUCCESS$_{AA}$ is true, then T and K2 should generally hold for CAN (because T and K2 are implied by SUCCESS$_{AA}$)

ii. T and K2 do not generally hold for CAN.
iii. Hence, SUCCESS$_{AA}$ is false.

In the remainder of this section, I am going to show that (i) is false.[13] Let "CAN" once more stand for the "can" of ability, and let it function as a sentences operator with scope over a proposition. Adapting T and K2 for CAN, we get the following two principles:

T*. S ϕ's → CAN(S ϕ's)
K2*. CAN(S ϕ's v S ψs) → CAN(S ϕ's) v CAN(S ψs)

Both, we know, fail to state laws. SUCCESS$_{AA}$ is perfectly in line with that. In fact, the truth of SUCCESS$_{AA}$ *explains why* T* and K2* fail to state laws.

On the one hand, it is not generally true, according to SUCCESS$_{AA}$, that if S ϕ's, S has the ability to ϕ. That is because it is not generally true that if S ϕ's, then S ϕ's in a sufficient proportion of the relevant possible situations in which S intends to ϕ. Take the general ability to hit the bull's eye and suppose all we hold fixed is the agent's amount of training (plus some background assumptions). Then obviously, the mere fact that I hit the bull's eye in actuality does not prove that I hit the bull's eye in a *sufficient proportion* of relevant possible situations in which I intend to hit it. It may clearly be the case that I fail miserably in almost all of the other situations in which I have had the same amount of training. In a nutshell: SUCCESS does not entail the general truth of T*, because SUCCESS yields that T* fails whenever "ability" is given a less than maximally specific interpretation.

SUCCESS$_{AA}$ does not entail the general truth of K2, either. By inserting the truth conditions SUCCESS$_{AA}$ postulates for ability statements into K2, we get:

K2*. If S ϕ's or ψ's in a sufficient proportion of the relevant possible situations in which S intends to ϕ or ψ, then S ϕ's in a sufficient proportion of the relevant possible situations in which S intends to ϕ, or S ψ's in a sufficient proportion of the relevant possible situations in which S intends to ψ.

This is not a valid principle either. It is, for instance, true that I draw a red card or a black card in a sufficient proportion of the relevant possible situations in which I intend to draw a red card or a black card. Actually, the proportion of

---

[13] Refuting (i) does not suffice to show, of course, that SUCCESS$_{AA}$ does not falsely entail the validity of any other invalid inferences. Since I cannot show, for any invalid inference, that SUCCESS does not entail its validity, however, I assume that refuting (i) is all that is required at this point.

such situations is presumably extremely high – there is not much that could stop me from realizing that kind of intention when faced with a stack of cards. It is not true, in contrast, that I draw a red card in a sufficient proportion of relevant possible situations in which I intend to draw a red card. My intention to draw red is not as closely linked to a successful performance. In fact, it will be no higher than 50 % – mere luck, given that there are only two colors available.

We can conclude that Kenny's counterexamples against possibilism do not spell any kind of trouble for the success view. On the one hand, the success view does not postulate that abilities are some kind of restricted possibilities. It is therefore not *per se* committed to the claim that the ability operator obeys the basic laws for possibility. On the other hand, there is no reason to assume that the success view is otherwise committed to the problematic principles. Instead, as I have shown, the truth conditions for ability statements given by SUCCESS$_{AA}$ account for the counterexamples against T* and K2* perfectly. If SUCCESS$_{AA}$ is true, then it rightly follows that T* and K2* are not generally true.

4.8 Finks and other unintentional abilities

In the last five sections, we saw that the success view is extremely powerful in that it accounts for all of the adequacy conditions for a comprehensive view of (agentive) abilities and does not run into the problems that proved destructive for the simple conditional analysis and possibilism. In this and the next section, I will discuss an objection to the success view that comes to mind quite easily. As it seems, there are cases in which an agent has an ability to ϕ, but can only exercise it if she does not intend to ϕ. Call abilities of that kind "unintentional abilities". In response to this objection, I will formulate a weaker version of the success view, according to which having an unintentional ability is for the agent to ϕ in a sufficient proportion of the relevant possible situations in which the agent intends to ψ, where "ψ" specifies a different action type than "ϕ", but ϕ-ing in response to intending to ψ counts as a success.

According to the success view, an agent has an agentive ability to ϕ if and only if the agent ϕ's in a sufficient proportion of relevant possible situations in which the agent intends to ϕ. But can we not easily think of cases in which an agent has an agentive ability, but can only exercise it if she does *not* intend to? In such cases, the success view yields wrong results. The modal success rate – the proportion of performance situations among the relevant intention situations – will be zero. The success view will therefore falsely predict that the agent lacks the ability in such cases.

The reader familiar with the debate on dispositions will presumably immediately think of cases that exhibit a similar structure: cases of finkish dispositions. Call a disposition *finkish* if and only if the occurrence of the stimulus interferes with the manifestation.[14] Many examples of finkish dispositions are quite far out – sorcerers and magic often play a crucial role (Lewis 1997); but in fact, we are surrounded by finkish dispositions in everyday life. An ordinary fuse, for instance, is a fink for a wire's disposition to conduct electricity when touched by a conductor (Molnar 1999). The wire has that disposition, but once it is attached to a fuse, touching the wire will make it go dead. Like in masking cases, the object will therefore not show the manifestation when the stimulus occurs. The reason, however, is not that something external interferes with the manifestation after the stimulus occurs (as in the case of masks), but rather that that the stimulus *itself* interferes with the manifestation.

Can we think of similar cases in the sphere of abilities? Let's call an ability *finkish* if and only if the intention itself interferes with the successful exercise of the ability. At first, it seems easy to think of abilities which meet this characterization. Some abilities wear their finkishness on their sleeves. Think of the ability to unintentionally read street signs;[15] or the ability to follow a conversation at another table without listening. Abilities of this kind can only be exercised when the agent does not intend to exercise them. That seems to be a problem for the success view. According to the success view, an agent has an ability to φ only if the agent φ's in a sufficient proportion of relevant possible situations in which the agent intends to φ. In cases in which the agent's intention interferes with her φ-ing, this condition will never be met. Hence, the success view seems to be unable to account for agents' finkish abilities.

We have to be careful here, though. The examples I just gave for finkish abilities are not very auspicious. The success view, as it stands, is an account of abilities to perform *actions*, recall. But reading street signs unintentionally and following a conversation at another table without listening are not actions. Rather, they are what I called "mere behaviors" (→ 1.4). Abilities to perform mere behaviors are what I called non-agentive abilities. Abilities of that kind are not properly captured in terms of a tie between intentions and performance. They call for a slightly different analysis. I will turn to non-agentive abilities in the next chapter.

---

14 In the case of finkish dispositions, the stimulus interferes with the manifestation *by making the disposition itself vanish* (Lewis 1997). I am using the term "finkish" more liberally here.
15 I owe this example to David Löwenstein.

Can we think of finkish *agentive* abilities? Differently put: are there *actions* which the agent has the ability to perform, but yet the intention to perform those very actions interferes with the successful performance? I think we can. Here are a few examples:

> CASUAL. Frieda has the ability to walk casually, but as soon as she intends to walk casually, she is so self-conscious that her walk becomes stiff and unnatural. Thus, it is not the case that she walks casually in a sufficient proportion of relevant possible situations in which she intends to walk causally.
>
> IN TUNE. Fred has the ability to sing in tune, but as soon as he intends to sing in tune, he becomes so strained that he inevitably sings out of tune. Thus, it is not the case that he sings in tune in a sufficient proportion of relevant possible situations in which he intends to sing in tune.
>
> WITTY. Emma has the ability to make witty and hilarious remarks in talks, but whenever she intends to make witty and hilarious remarks, she overdoes it and tends to create awkward situations. Thus, it is not the case that she makes witty and hilarious remarks in a sufficient proportion of relevant possible situations in which she intends to make witty and hilarious remarks.

The structure of these cases is as follows. There is something an agent has an ability to do, but the intention to do the very thing gives rise to some kind of impediment to actually doing the thing in question. Frieda becomes self-conscious, Fred becomes strained, Emma overdoes the joking. The intention itself results in an impediment to successful performance. What the agents are able to do are obviously actions: walking casually, singing in tune, and making witty and hilarious remarks. Hence, they are not excluded by the explicit restriction of SUCCESS$_{AA}$ to agentive abilities.

Apparently, then, there are agentive abilities to ϕ which do not require the agent's performances of ϕ-ing to be appropriately tied to the intention to ϕ. Sometimes, agents have an ability to ϕ and yet fail miserably in situations in which they intend to do the thing in question. I think this is just the way things stand. The cases do indeed pose counterexamples to the success view as I have formulated it. In finkish cases, there will not be a sufficient proportion of performance cases among the relevant intention situations.

It is worth noting that finkish abilities are only a special case of a much broader range of counterexamples to the success view as it stands. In the finkish case, the intention *itself* interferes with the successful performance of ϕ-ing. But this need not be so. To come up with counterexamples to SUCCESS, there may just as well be an impediment to the agent's successful performance of ϕ-ing upon intending to ϕ, which is completely independent from the agent's forming the intention to ϕ itself. Here are two cases of that kind:

HANDSTAND. When Ela intends to do a handstand against a wall, she always fails, because she doesn't have the slightest idea of how to do it. But when she intends to put her hands on the ground, throw up her legs and hit her feet through the wall, she ends up doing a perfect handstand.[16]

POUR ÉLISE. Peter has the ability to play "Pour Élise" without any mistakes. *A forteriori*, he has the ability to play the middle part of "Pour Élise" without any mistakes. However, he never succeeds in playing the middle part without mistakes upon intending to. He can only play the piece, and thus the middle part, flawlessly in one go – start to finish.[17]

In both cases, the agent has an ability to perform an action, yet the agent does not exhibit the modal tie between intentions and performances postulated by $SUCCESS_{AA}$. It is not the case that there is a sufficient proportion of ϕ-ing situations among the relevant possible situations in which the agent intends to ϕ. Yet, the reason for the agents' respective failures to meet $SUCCESS_{AA}$ has to do with impediments that are fully independent from the forming of the intention to ϕ itself. In HANDSTAND, the problem is that Ela does not know enough about the proper method of doing a handstand. In POUR ÉLISE, the problem is that whatever program in Peter's brain underlies his piano playing ability, this program can only run through the whole piece flawlessly, but not through parts. In neither case does the intention to ϕ itself cause the trouble.

Finkish abilities are a special instance of a whole range of cases in which an agent has an ability which can only be exercised as long as the agent does not intend to exercise it, then. Let us call such abilities "unintentional abilities". Unintentional abilities obviously pose a problem for the success view.

Luckily, unintentional abilities can be dealt with. Despite their obvious impact on the success view, unintentional abilities merely show that the view has to be extended a bit. Thus far, the success view says that for an agent to have an ability to ϕ, the agent has to ϕ in a sufficient proportion of relevant possible situations in which she intends to ϕ. That is obviously not the case in the examples above. The agent's ϕ-ing is not appropriately tied to her intention to ϕ.

There is, however, a *different* intention to which each agent's performance of ϕ-ing is appropriately tied, even in the case of unintentional abilities. In the finkish cases, Frieda walks casually in a sufficient proportion of the relevant possible situations in which she intends to walk, for instance. Fred sings in tune in a sufficient proportion of the relevant possible situations in which he intends to sing. Emma make witty and hilarious remarks in a sufficient proportion of the relevant possible situations in which she intends to make remarks.

---

[16] The case has been brought to my attention by Wolfgang Schwarz (personal conversation).
[17] The case has been brought to my attention by Barbara Vetter (personal conversation).

In the non-finkish cases, the same structure can be observed. Ela does a handstand in a sufficient proportion of the relevant possible situations in which she intends to hit her feet against the wall. Peter plays the middle part of "Pour Élise" flawlessly in a sufficient proportion of the relevant possible situations in which he intends to play "Pour Élise".

This suggests the following modification of the success view. On a more liberal understanding of the modal tie between intentions and performances that is required for an agent to have an ability to ϕ,

> SUCCESS$_{AA}$*. an agent S has an agentive ability to ϕ if and only if there is *some* intention ψ, such that S ϕ's in a sufficient proportion of similar situations in which she intends to ψ, where ϕ and ψ need not specify the same action type.

This formulation is more liberal than the original formulation in that the intention to which the agent's performances of ϕ-ing have to be modally tied can (and as a rule, will be), but need not, be the intention to ϕ itself. It may well be some other intention.

Is the modification *ad hoc*? It is not. Instead, SUCCESS$_{AA}$ and SUCCESS$_{AA}$* simply account for two varyingly strong senses of what it is for an agent to have an ability. There seems to be an important difference between an understanding of abilities according to which the agent's ϕ-ing has to be appropriately tied to the intention to ϕ itself on the one hand, and an understanding of abilities according to which the agent's ϕ-ing merely has to be tied to *some* intention, on the other.

In the former, but not in the latter case, the agent is able perform the act in question *at will* and is able to perform it *upon request*. Usually, I should think, this is the sense of ability we are interested in when we are thinking about abilities to perform actions. This sense is captured by the original success condition SUCCESS$_{AA}$. Very importantly, it is a sense in which the agents in the examples lack the ability to ϕ. Frieda cannot walk casually upon request; Fred cannot sing in tune upon request; Emma cannot make witty and hilarious remarks, Ela cannot do a handstand, and Peter cannot play the middle part of "Pour Élise" flawlessly upon request.

Cases of unintentional abilities – cases in which the performance of the action is properly linked, not to the intention to perform that very action, but some other intention – seem to be deficient in one very crucial respect, then. There is a somewhat accidental element to abilities of that kind. Frieda's walk tends to *turn out* to be casual, Fred's singing tends to *turn out* to be in tune, and Emma's remarks tend to *turn out* to be funny. Likewise, Emma's attempt to hit her feet against the wall tends to result in a handstand, and Peter's play of "Pour

Élise" tends to involve a flawless performance of the middle part. But this particular feature of their respective actions is not something the agents bring about intentionally – it is not fully controlled by the agents.[18]

This suggests the following picture: whenever an agent has an agentive ability, there is *some* action which the agent can perform upon request – there is an action which is appropriately linked to the agent's intention to do that very thing. In the examples, the actions of that kind are walking, singing, giving talks, performing certain bodily movements, and playing "Pour Élise" start to finish. For those actions, the original success condition $SUCCESS_{AA}$ holds.

In some cases, however, the ability we ascribe is not the ability to do that very thing, but a different thing: walking casually, singing in tune, making hilarious remarks, doing a handstand, and playing the middle part of "Pour Élise" flawlessly, for instance. Why do we do that? Because the thing the agent is able to do upon request is such that it usually turns out in a certain way: casually, in tune, and so forth. That is why the agent's walking casually, singing in tune, and making witty and hilarious remarks are themselves modally tied in the right way to the intention to do whatever the agent can do upon request. The original formulation of the success condition is still the primary one on this picture. But the more liberal formulation comes into play whenever there is something the agent can do, even though she cannot do it upon request or at will.

End of story? Not quite. One may object that $SUCCESS_{AA}{}^*$ is in fact a bit *too* liberal. On my account, an agent has an ability to φ, in the liberal sense, whenever there is some intention, no matter whether it is the intention to φ or not, to which the agent's φ-ing is closely enough tied. But that does not seem right, does it? Suppose I have the disposition to twitch my left eye whenever I am nervous, and to feel nervous whenever I talk to Ben, which in turn I do pretty much whenever I intend to. It seems as though $SUCCESS_{AA}{}^*$ yields that I have the ability to twitch my left eye. There is some intention – the intention to talk to Ben – such that I twitch my left eye in a sufficient proportion of relevant possible situations in which I form that intention. And that is weird. It does not seem as though I have an ability to twitch my left eye. My eye twitching is just something that happens to me when I am nervous. I have no control over it whatsoever.

---

18 That there is a lack of control might be least obvious in Peter's case of playing the middle part of "Pour Élise". But what actually happens in that case, I take it, is that once Peter starts playing the piece, some automatism sets in, which carries him flawlessly through the whole piece. Peter does not, in that sense, fully control what is going on. He is being carried along a great deal. Of course, there is some control involved: the agent can stop playing the middle part. But there is not full control over what is happening and how.

Again, we have to be careful here, though. Recall that we are dealing exclusively with agentive abilities. Agentive abilities are abilities to perform actions. And twitching my eye fails to be just that: it is not an action. Hence, there is indeed one restriction to SUCCESS$_{AA}$*, which I should once more add very explicitly: φ-ing has to be an action. It must not be a mere behavior.

Are there still counterexamples to SUCCESS$_{AA}$*? I find it hard to think of one, but perhaps there are some. If you can think of one, let me put you on hold until the next chapter. For as we'll see there, SUCCESS$_{AA}$* is not fully spelled out quite yet. What is missing, and what is actually quite crucial to the success view, is the requirement that there has to be success involved in the performance of φ-ing. More specifically, SUCCESS$_{AA}$* misses out on the fact that in addition to the agent's φ-ing in a sufficient proportion of the relevant possible situations in which the agent has some intention, the agent's φ-ing in response to that intention also has to count as a success. Really, then, SUCCESS$_{AA}$* needs to be replaced by the following, more subtle view:

> SUCCESS$_{AA\_COMPLETE}$. An agent S has an agentive ability to φ if and only if (i) there is *some* intention ψ, such that S φ's in a sufficient proportion of similar situations in which she intends to ψ, where φ and ψ need not specify the same action type, as long as (ii) φ-ing in response to ψ counts as a success.

Condition (ii) may not be fully accessible at this point, but things will fall into place in the next chapter. For now, let us just note that there is a more liberal notion of having the ability to φ, which has to do with the fact that what the agent intends to do need not be precisely what she ends up doing. In the next section, I'll differentiate this insight further.

4.9 Impairments by ignorance

The cases in the last section seemed to indicate that agentive abilities do not require a proper link between the intention to φ and the agent's φ-ing. That is because in cases of finkish abilities and unintentional abilities more generally that link is violated. Yet, the agent has the ability. I argued that unintentionally abilities are very interesting, but not actually problematic for the success view. What they show is that there are actually two senses of an agent being able to do something, which should be kept separate. The sense in which the agent has the ability in cases of unintentional abilities is weaker than the standard sense of ability, but a proper sense nevertheless. What I established in the last section is a distinction between two varyingly strong readings of the success condition, which provides the resources to distinguish systematically between those two senses.

In this section, we'll look at a further kind of unintentional agentive abilities: cases that involve what I will call *impairments by ignorance*. Here are some cases illustrating this phenomenon:

> C MINOR. Klaus has the ability to grab a C minor chord. But he doesn't know that the chord he so effortlessly grabs and thinks of as "the cool chord" is a C minor. In fact, he thinks "C minor" is the name for the D minor chord. Thus, it is not the case that he grabs a C minor in a sufficient proportion of relevant possible situations in which he intends to grab a C minor.
>
> CHATTING. Karl has the ability to chat casually with the president of the United States, but only in virtue of the fact that he fails to recognize him. Knowingly talking to the president would actually freak him out. Thus, it is not the case that he chats casually to the president in a sufficient proportion of relevant possible situations in which he intends to chat casually with the president.
>
> BRESLAU. Ela has the ability to cycle from Warsaw to Breslau. Ela believes, however, that "Breslau" is the German word, not for the Polish city Wroclaw, but for the country of Belgium. Hence, it is not the case that Ela cycles to Breslau in a sufficient proportion of relevant possible situations in which she intends to do so. She would either end up in Belgium, or, even more likely, give up somewhere along the way.

Again, these cases seem to show that the link between intentions and performances that is postulated by the success view is not necessary. The agents in the examples have an ability to φ, while it is not the case that their φ-ing is properly linked to their intentions to φ.

Against the background of the last section, one response to the examples is obvious. As in the last section, the agents clearly meet SUCCESS_{AA}*. There is *some* intention such that S φ's in a sufficient proportion of situations in which she forms *that* intention[19]. Klaus grabs a C minor in a sufficient proportion of similar situations in which he intends to grab "the cool chord". Karl chats casually with the president in a sufficient proportion of similar situations in which he intends to chat with "that guy" (where "that guy" refers to the president). Ela cycles to Breslau in a sufficient proportion of similar situations in which she intends to cycle to Wroclaw.

What this shows is that all of the agents have the ability to φ in the liberal sense established in the last section: for all of them, there is some intention they can form, such that they φ in a sufficient proportion of similar situations in which they form *that* intention. What they fail to exhibit is the ability in the

---

[19] And doing so counts as a success, but let's postpone this until the next chapter.

stronger, original sense: the intention that is properly tied to their ɸ-ing is not the intention to ɸ.

Yet, there is more to say about cases of impairments by ignorance. To see that, note that there seems to be a relevant difference between the sense in which the agents we considered in the last section have the ability to do something and the sense in which the agents we are considering right now have an ability. In the last section, the cases we looked at were crucially characterized by the fact that there were certain parts or aspects of the agents' doings that are not fully controlled by the agents. Our agents were able to walk, sing, try to hit their feet against the wall, and to play "Pour Élise" in the strong sense, while not being in the same strong sense able to walk *casually*, sing *in tune*, do a handstand, or play the middle part of "Pour Élise". These aspects of parts of their doings were something they did not fully control. In that sense, we said, our agents are only weakly able to do the things in question.

In cases of impairment by ignorance, diminished control is not at issue at all. The agents control their actions fully when exercising them. There is no passive or automatized element to their doings whatsoever. In this respect, they seem to resemble agents who are able to perform an action in the strong sense. In terms of control, they resemble agents whose intention *to ɸ*, and not just some other intention, is properly linked to their performances of ɸ-ing.

Why is that? The answer has to do with the fact that in cases of impairments by ignorance, there is a much closer relation between the intention to which the agent's performances of ɸ-ing are tied and the act of ɸ-ing itself than in other cases of unintentional abilities. Cases of impairment by ignorance differ from other cases of unintentional abilities in one very crucial respect. The thing the agents intend to do *is* the act of ɸ-ing, albeit under a different description. Grabbing "the cool chord" *is* grabbing C minor. Chatting casually to "that guy" *is* chatting casually to the president. Cycling to "Breslau" *is* cycling to Wroclaw.

The same is not true in the cases we looked at in the last section. It is not true that just walking is walking casually, that just singing is singing in tune, or just trying to kick one's feet through the wall is doing a handstand. It is this difference between the cases of impairments by ignorance and other unintentional abilities which accounts for the fact that ignorant agents, while resembling the agents from the last section in not exhibiting the proper tie between their intentions to ɸ and their performances of ɸ-ing, have full control over their performances of ɸ-ing.

This yields the following picture: in all cases of unintentional abilities, agents are able to ɸ only in the liberal sense established in the last section. They are able to ɸ only in the sense that there is some intention, the intention to ψ, such that the agents ɸ in a sufficient proportion of the relevant possible

4.9 Impairments by ignorance

situations in which they form *that* intention. But while ψ and φ are not necessarily closely related in unintentional abilities in general, they pick out one and the same action in the cases of ignorance. This accounts for the fact that there is no lack of control involved in the latter.

The basic insight we need in order to be able to understand how impairments by ignorance differ from the cases in the last section, then, is that "to intend to φ" can be read in two different ways: it can be given a *de dicto* reading or a *de re* reading (*locus classicus:* Quine 1956). In the *de dicto* reading, one intends to φ only if the content of one's intention is "to φ". In the *de re* reading, in contrast, one also intends to φ if the content of one's intention is "to ψ", where ψ-ing is φ-ing.

The objection from impairments by ignorance obviously depends on a *de dicto* interpretation of "intending to φ". In the examples, it is not the case that the agent φ's in a sufficient proportion of the relevant possible situations in which the agent intends, *de dicto*, to φ. Klaus, for instance, grabs D minor whenever he intends, *de dicto*, to grab C minor. However, and this is the important point, in the *de re* sense of "intending to φ", the agents' intentions *are* properly tied to their performances. Klaus grabs C minor in a sufficient proportion of the relevant possible situations in which he intends, *de dicto*, to grab the cool chord. But since grabbing the cool chord just is grabbing C minor, it is also the case that he grabs C minor in a sufficient proportion of the relevant possible situations in which he intends, *de re*, to grab C minor. Likewise in the other cases.

SUCCESS$_{AA}$ can be read in two ways, then:

(1) An agent S has the ability to φ if and only if S φ's in a sufficient proportion of the relevant possible situations in which S intends, *de dicto*, to φ.

(2) An agent S has the ability to φ if and only if S φ's in a sufficient proportion of the relevant possible situations in which S intends, *de re*, to φ.

In cases of impairments by ignorance, (1) is violated. But (2) is not. And thus, there is a good sense in which the agents have the ability to φ in the strong sense spelled out by SUCCESS$_{AA}$, and not just in the weak sense spelled out by SUCCESS$_{AA}$*, in those cases. Cases of impairments by ignorance are therefore unintentional only on the *de dicto* reading of SUCCESS$_{AA}$.

4.10 Upshot

Let's wrap up our findings in this chapter. The objective of the chapter was to develop a comprehensive view of agentive abilities – abilities to perform actions. I suggested, somewhat provisionally, that having an agentive ability is a matter of modal success in the sense that the agent has to φ in a sufficient proportion of the relevant possible situations in which the agent intends to φ. I called this "the success view of agentive abilities", or SUCCESS$_{AA}$ for short.

SUCCESS$_{AA}$ is a hybrid of the simple conditional analysis and possibilism, in the sense that SUCCESS$_{AA}$ combines both the idea that abilities are a matter of a modal tie between the agent's intentions and performances and the idea that abilities are always had in view of certain facts, where those facts impose restrictions of the realm of the possible situations.

After introducing the basic framework, I first (→ 4.2) elaborated on the notion of an intention, as it figures in the success view, and provided a rationale for formulating the view in those terms. An intention, I said, is simply an *action initiating* propositional attitude in the sense that it is part of its causal role that it will typically initiate behavioral episodes corresponding to its content. The behavioral episodes which are caused by intentions (in the right way)[20] are actions. Why does the success view feature intentions and not any other motivational state? Because only intentions are both action-initiating in the specified sense and not themselves actions. This is important, because only an action-initiating state secures that a failure to φ really counts against the ability and does not simply trace back to overriding other mental attitudes, and only a mental state that is not itself an action secures that SUCCESS$_{AA}$ does not run into a regress.

The lion's share of the chapter (→ 4.3–4.7) was devoted to presenting the merits of SUCCESS$_{AA}$. As I argued at great length, SUCCESS$_{AA}$ accounts for all of the adequacy conditions for a comprehensive view of (agentive) abilities and does not run into any of the problems that beset the simple conditional analysis and possibilism. Let me briefly walk you through those merits once more.

In section 4.3, I showed that the success view provides a comprehensive account of degrees and the corresponding kind of context sensitivity of ability statements. The degree of an ability to φ, I argued, corresponds to the weighted proportion of the relevant intention situations in which the agent φ's. Here, we

---

[20] "In the right way" means: by non-deviant causal chains. How to think of non-deviance is a problem of action theory and falls outside the scope of this book. For an overview see Wilson & Shpall (2012, section 2).

added the first important supplement to the original statement of the success view. What matters, for an agent's ability to ϕ is that the agent ϕ's in a sufficient *weighted* proportion of the relevant possible situations in which the agent intends to ϕ.

In section 4.4, I showed that the success view does not run into problems with cases of impeded intentions. The success view entails what I called "the existential requirement": for SUCCESS$_{AA}$ to be met, there have to be relevant possible intention situations to begin with. As I argued, the existential requirement is not met in cases of impeded intentions. In such cases, there are neither relevant intention situations, nor are there any performance cases among the relevant intention situations. That is not to say that there is *no* good sense in which coma patients retain their ability to raise their arm whatsoever: depending on whether the coma is varied or held fixed across the possible situations, there will or will not be relevant intention situations and arm raising situations among those. This is exactly right: the problem of the impeded intention only arises when we are interested in what the agent can do *in view of the impediment*, such as the coma. When we are interested in what the agent can do, abstracting away from the coma, the agent retains lots of abilities. The success view can account smoothly for both senses.

In section 4.5, I argued that the success view provides a comprehensive account of general and specific abilities. When thinking about an agent's general abilities, we hold different sets of facts fixed than when we think about her specific abilities. In the case of general abilities, we are interested in what the agent can do in view of her stable, intrinsic properties. In the case of specific abilities, we are interested in what she can do in view of all kinds of temporary, extrinsic features that concern her current situation.

In the course of the section, I also systematized various ideas about the general/specific distinction that have been floating around in the literature. In section (a), I looked more closely at general abilities and argued against the idea that we only ever hold intrinsic features fixed in the case of general abilities. Instead, I argued, stability of features is what counts. I also argued that there is not one sense of "general ability", but many, which vary in accordance to just which set of stable, mostly intrinsic properties we hold fixed.

In section (b), I looked more closely at specific abilities and argued that there is not one sense of "specific ability", but many. I explicitly distinguished three that seem to me to be of particular systematic interest: (i) particular abilities, or abilities that are had in view of the totality of facts that obtain at some point in time, *including* the agent's motivational states; (ii) conative abilities, or abilities that are had in view of all facts of a situation *except for* the agent's motivational states; and (iii) opportunities, or abilities that are had in view of the features of the situation that are external to the agent. I also argued that there are many more good senses of "specific ability", which vary in accordance to just which set of features of the situation we hold fixed.

In section (c), I looked at the ways in which general and specific abilities depend on one another. Here, I argued that some specific abilities will be such that having such an ability requires having a corresponding general ability. I also argued that having a general ability generally requires that there be various possible situations in which the agent has a corresponding specific ability.

In section (d), I looked at degrees of specificity and argued that an ability A1 is more specific than an ability A2 if and only if the kinds of facts that are held fixed in the case of A1 are a proper subset of the kinds of facts that are held fixed in the case of A2. I also argued that this only yields a partial ordering of the specificity of abilities, which I take to be exactly right.

In section (e), finally, I looked at the ways in which hypothetical circumstances enter our thinking about abilities. I argued that the structure of ability statements has four placeholders for circumstances: "S can φ" is to be read as "In C1, S can, in view of C* and when C3, φ-in-C2", where hypothetical circumstances can enter via C1, C2, and C3, yielding different interpretations of the sentence.

In section 4.6, we turned to masks and saw that the success view explains the workings of masks very smoothly. Masks are not a problem for the success view, because the mask – the interfering factor in the actual situation – is usually varied across the relevant possible situations. That means that the very small proportion of situations in which the mask is present do not in any sense have priority over situations in which it is not present. And hence, the fact that the actual situation is a mask situation is not of particular interest. The actual situation does not have any special status among the realm of the relevant possible situations. Despite the mask, there may well be a sufficient proportion of performance cases among the relevant intention situations.

In section 4.7, I showed that the formal problem Kenny noted about possibilism does not arise for the success view. Quite to the contrary: the success view explains very smoothly why the laws that hold for the "can" of possibility do not hold for the "can" of ability.

In sections 4.8 and 4.9, finally, we turned to two closely related natural objections to the success view. In section 4.8, I discussed the objection that the success view runs into problems with what I called "unintentional agentive abilities" – agentive abilities which can only be exercised as long as the agent does not intend to exercise them. In section 4.9, the worry was refined by examining a particular kind of unintentional agentive abilities: abilities that can only be exercised as long as the agent does not intend to exercise them, because the agent is epistemically impaired in certain ways.

In response to these objections, I introduced a somewhat weaker version of the success view, which accommodates such cases: on that view, an agent has an agentive ability to φ if and only if there is some intention to ψ, such that the agent φ's in a sufficient proportion of the relevant possible situations in which the agent intends to ψ, where φ and ψ need not specify the same action type, but where φ-ing in response to intending to φ counts as a success. Depending on whether ψ and φ pick out the same action type and depending on whether they pick out the same action type *de dicto* or *de re*, the view yields varying senses of "having an ability".

We are now in a position to formulate the fully comprehensive view of agents' agentive abilities we were after in this chapter:

> AGENTIVE ABILITIES. An agent S has an agentive ability to φ if and only if S φ's in a sufficient proportion of the relevant possible situations in which S intends to ψ, where ψ will typically be *de dicto* or *de re* identical to φ and in any case be such that φ-ing in response to the intention to ψ counts as a success.

As I said: the last bit of this condition – that φ-ing in response to the intention to ψ has to count as a success – will strike you as mysterious, because so far I have not talked about success as such. It's a cliff hanger. Everything will make sense in the next chapter, where we will proceed to extend the view to abilities of the non-agentive kind.

5 The success view II – Non-agentive abilities

So far, I have focused exclusively on agentive abilities – abilities to perform actions. In this chapter, let me turn to non-agentive abilities: abilities to engage in behaviors that are not actions and that are typically, or can even *only* be exercised without a foregoing intention on the agent's part. Examples of non-agentive abilities include biological abilities like the ability to digest, produce saliva or adapt one's pupil size to the light conditions, perceptive abilities like the ability to hear, see, or smell, and overtly non-agentive abilities like the ability to unintentionally read street signs or advertisements.[1]

Non-agentive abilities elude the success view as I have formulated it for agentive abilities. Like the simple conditional analysis, the success view of agentive abilities postulates a modal tie between intentions and performances. When it comes to non-agentive abilities, however, intentions do not seem to play a crucial role. And thus, it seems odd to analyze non-agentive abilities in terms of a modal tie in which intentions are so crucially involved. The suspicion suggests itself that the success view, while reasonable for agentive abilities, reaches its limits when non-agentive abilities come into focus.

As we will see in the course of this chapter, the success view of agentive abilities is indeed ill-suited to account for abilities of the non-agentive kind. On closer inspection, however, it will become clear that there is a version of the success view that applies to both agentive and non-agentive abilities alike.

As I'll argue, abilities are quite generally a matter of success across a sufficient portion of modal space. In the case of agentive abilities, that is to say that S has an ability if and only if S φ's in a sufficient proportion of the relevant possible situations in which the agent forms some suitable intention. In the case of non-agentive abilities, a structurally analogous analysis can be provided. On the view I'll lay out in what follows, S has a non-agentive ability if and only if S φ's in a sufficient proportion of the relevant possible situations in which some S-trigger for φ-ing is present, where an S-trigger is a trigger for φ-ing, in response to which φ-ing is a success.

From here, one can go in two directions. One can leave it at that and argue that non-agentive abilities are abilities insofar as they exhibit an analogous modal structure to agentive abilities. As far as one is willing to concede that intentions are S-triggers for actions, however, one can go even further and formulate a unified set of truth conditions for abilities *tout court*. On that view, S has an

---

[1] Since this example will come up so frequently in this chapter, I should once again note that I am borrowing it from Löwenstein (2017).

ability *tout court* if and only if S ϕ's in a sufficient proportion of the relevant possible situations in which some S-trigger for ϕ-ing is present. Combined with the idea that intentions *are* S-triggers for actions, this yields that agentive abilities are special instances of abilities *tout court*. Both agentive and non-agentive abilities fall under the same analysis, on this view, the only difference being that the S-triggers for actions and mere behaviors differ. I myself have sympathies with the unified picture. But one can go either way here.

5.1 The success view for non-agentive abilities

Abilities, I said in the beginning of the last chapter, are a matter of success. In the case of agentive abilities, this basic idea was spelled out in terms of what I then called the agent's modal success rate: the proportion of performance cases among the relevant intention situations. Apparently, this account does not carry over to unintentional abilities one-to-one.

What *does* carry over one-to-one, however, are the structural features of the account. Non-agentive abilities, too, are a matter of success. As in the case of agentive abilities, this idea can be spelled out in terms of a modal success rate. Again, what matters is the proportion of performance cases among a subset of the relevant possible situations. The only difference is the subset which matters. In the case of agentive abilities, the agent's intention plays a crucial role. Success, in the case of actions, has to do with a sufficiently strong modal tie between intentions and performances. That is why the relevant possible situations are restricted to the ones in which the agent forms a proper intention.

In the case of mere behaviors (which is the word I introduced for behaviors that are not actions), such as hearing, digesting, unintentionally reading street signs, or dissociating, in contrast, the agent's intention does not play a crucial role. Instead, in such cases, success consists in showing the right response to certain external triggers in the world, so to speak. Hearing sound when there is sound is a success. Digesting food when food is ingested is a success. Unintentionally reading a street sign when facing a street sign is a success. Dissociating when subjected to a potentially traumatic experience is a success.

Structurally, however, non-agentive abilities are not all that different from their agentive siblings. Performing an *action* of ϕ-ing is a success when exercised in response to an intention to ϕ. Being engaged in a behavior that is *not* an action is a success when exercised in response to some suitable external trigger. But apart from that, I submit, the analysis of the two runs along the very same lines. Thus, on the view I want to suggest,

SUCCESS$_{NAA}$. an agent S has a *non-agentive* ability to be engaged in some mere behavior ϕ if and only if S ϕ's in a sufficient proportion of the relevant possible situations in which some S-trigger for ϕ-ing is present.

An S-trigger for ϕ-ing, as I would like to use that word, is any trigger such that ϕ-ing in response to that trigger counts as a success.[2] Note that S-triggers play the same functional role in SUCCESS$_{NAA}$ that intentions play in SUCCESS$_{AA}$. They mark out a feature of a situation that has to be present in order for ϕ-ing in that situation to be a success. The only difference is that we can specify what feature of the situation that is in the case of agentive abilities: the situation has to be one in which a suitable intention is formed.

Why? Because for an action to be a success is precisely for the action to follow a suitable intention. In the case of non-agentive abilities, we cannot specify the feature of the situation in a similar vein. All we can say is that there has to be some trigger, such that ϕ-ing in response to it is a success. Really, though, the role of that is no different from the role of intentions in SUCCESS$_{AA}$. I'll talk more about the exact relation between SUCCESS$_{AA}$ and SUCCESS$_{NAA}$ later on (→ 5.3). For now, let's focus exclusively on SUCCESS$_{NAA}$.

Let's go through some of the examples of non-agentive abilities to see how SUCCESS$_{NAA}$ applies. First, abilities to have sense perceptions. An agent has the non-agentive ability to smell, on the view I have just laid out, if and only if the agent smells in a sufficient proportion of relevant possible situations in which some S-trigger for smelling is present. What is an S-trigger for smelling? Subjection to odor, I guess. Thus, an agent has the ability to smell if and only if she smells in a sufficient proportion of relevant possible situations in which she is subjected to odor.

The scheme is the same in the case of all perceptual abilities. In the case of seeing, substitute "subjection to odor" with "subjection to some visual stimulus"; in the case of hearing, substitute it with "subjection to sound"; in the case of tasting, substitute it with "subjection to some gustatory stimulus".

What about abilities to undergo biological processes? Take the ability to sweat. What is an S-trigger for sweating? That the agent's body needs cooling, I guess. Thus, SUCCESS$_{NAA}$ yields that an agent has the ability to sweat if and only if the agent sweats in a sufficient proportion of relevant possible situations in which the agent's body needs cooling.

Likewise in the case of other biological abilities. In the case of adjustment of the pupil size, the S-trigger is the agent's subjection to a change in the lighting

---

[2] If you don't like the causal connotation of "in response to", please feel free to substitute it with a less causal-sounding term like "vis-à-vis".

conditions. In the case of producing saliva, the S-trigger is that the agent's body needs saliva.

The same, finally, goes for the examples of overtly non-agentive abilities – abilities which are unintentional for conceptual reasons. To evaluate whether an agent has the ability to unintentionally read street signs, we have to check whether she reads street signs in a sufficient proportion of the relevant possible situations in which some S-trigger for unintentionally reading street signs is present. What could be such a trigger? One's subjection to street signs, I should think. Thus, an agent has an ability to unintentionally read street signs if and only if she unintentionally reads street signs in a sufficient proportion of the relevant possible situations in which she is subjected to street signs.

This, then, is the success view for unintentional abilities. In the following sections, it is going to be worked out in more detail.

5.2 Degrees and the distinction between general and specific abilities

The success view for non-agentive abilities resembles the success view of agentive abilities closely (more on the exact relation between the two in the next section), and in fact, it functions analogously in all crucial respects. As before, the domain of the relevant possible situations will vary across contexts, which accounts for the distinction between general and specific non-agentive abilities. Moreover, the degree of an agent's non-agentive ability to φ will be determined in the same way as the degree of an agent's agentive ability to φ: it is determined by the success rate. Thus, in the case of non-agentive abilities, it is given by the proportion of the performance cases among the relevant possible S-trigger situations.

Let me briefly elaborate on both of those aspects of the view. First, degrees. Like agentive abilities, non-agentive abilities come in degrees. A child does not have the ability to unintentionally read street signs at all; reading street signs, or reading anything at all, for that matter, requires effort and is hard work. The better the reading skills of a reader become, however, the more often will it happen that the agent grasps the meaning of written words just by glancing over them. The agent has become much better at unintentionally reading street signs. Eventually reading short words and phrases becomes an automatism. The reader is now very good at unintentionally reading street signs.

The success view for non-agentive abilities accounts for the stages of an agent's ability to unintentionally read street signs in terms of a variance in the agent's success rate. The better an agent is at some doing φ, the higher the proportion of ϕ-ing cases among the relevant S-trigger situations. The child who is unable to ever read anything unintentionally will not read any street signs unintentionally in situations in which her amount of training in reading is held constant and she is faced with a street sign. The more practiced reader will unintentionally read quite a few street signs already in situations in which her amount of training is held constant and she is faced with a street sign. The highly trained reader, finally, will unintentionally read virtually any street sign in situations in which her amount of training is as in actuality and she is faced with a street sign.

So far, we have gotten along with the simple proportion of performance cases among the relevant possible trigger situations to account for the variances we observe in agents' degrees of non-agentive abilities. All we had to do, so far, is count. What this shows is that an agent's ability to unintentionally read street signs can simply be evaluated on the dimension of reliability. In the case of agentive abilities, we saw that there is usually a second dimension involved: the dimension of achievement. How good an agent is at ϕ-ing has to do, in part, with how *well* she performs in various situations.

When it comes to non-agentive abilities, this dimension can play a role as well. Take perceptive abilities, like the ability to see. How good someone is at seeing depends not only on the range of situations in which the agent sees. Instead it depends to large parts on how *well* the agent is seeing in those situations: how clearly she sees things, say. Or take the psychological ability to dissociate. How good someone is at dissociating will not only depend on how reliably the person dissociates in traumatic situations. It will also depend, for instance, on how fully she dissociates. How fully, that is, she manages to encapsulate the traumatic experience.

As in the case of agentive abilities, the degree of an agent's non-agentive ability will hinge not only on the simple proportion of performance situations among the relevant S-trigger situations, but also on the agent's degree of achievement in each of the performance cases. What we need to do, then, is to weigh the performance cases in accordance to the agent's degree of achievement. The overall degree of an agent's ability to ϕ will eventually hinge on one or both of those dimensions, depending on context. All of this is old news. The success view yields the same account of degrees in the case of non-agentive abilities as in the case of agentive abilities that I suggested earlier on (→ 4.3).

Let's move on to the distinction between general and specific non-agentive abilities. I take it to be clear that there is no difference between agentive and non-agentive abilities when it comes to the need for the distinction between general and specific abilities. Non-agentive abilities, too, can be general or specific, and the two come apart in a range of cases. I can have the general ability to unintentionally read street signs, for instance, but lack the specific ability to unintentionally read street signs in a particular situation.

The agent may, for instance, be blindfolded. She may have taken some drug which impairs her automatic recognition mechanisms. Or there may simply be no street signs around for her to read. In all of those cases, the agent is unable to unintentionally read street signs in the particular situation she finds herself in. Yet, the blindfold, the drug, and the absence of street signs do not impair her general ability to unintentionally read street signs.

The success view for non-agentive abilities yields the right results in those cases because the sets of the relevant possible situations will differ depending on the kind of ability we have in mind. When we are interested in the agent's specific abilities, the agent's being blindfolded, drugged, or not in sight of any street signs will be held fixed across the possible situations. The relevant possible situations will contain only situations in which the agent has the mentioned (intrinsic or extrinsic) properties. And given that selection of relevant possible situations, it will not be the case that the agent reads street signs unintentionally in a sufficient proportion of the relevant intention situations.

When we are considering the agent's general ability, in contrast, we are not holding those facts fixed across modal space. All we are holding fixed is the agent's automatic recognition mechanism plus some background assumptions about the world, for instance. And given *that* selection of the relevant possible situations, there *will* be a sufficient proportion of relevant intention situations in which the agent unintentionally reads street signs.

5.3 The broader scheme: the success view of abilities tout court

Let's take stock. We started out, in the last chapter, with an account of agentive abilities. On the preliminary version of that view, which we worked with throughout the largest part of the last chapter,

> SUCCESS$_{AA}$. an agent S has an agentive ability to perform an action ϕ if and only if S ϕ's in a sufficient proportion of the relevant possible situations in which S intends to ϕ.

In this chapter, we looked at non-agentive abilities – abilities to be engaged in doings which are not actions. Such mere behaviors, as I called them, can or can *only* be exercised without a foregoing intention on the agent's part. For abilities of that kind I suggested a structurally analogous analysis. The only difference between the two analyses is that the restriction to intention situations, which played a crucial role in connection with agentive abilities, has been swapped for a restriction to situations in which an S-trigger for ϕ-ing is present – a trigger in response to which ϕ-ing counts as a success. Thus, I said,

> SUCCESS$_{NAA}$. an agent S has an *non-agentive* ability to be engaged in some mere behavior ϕ if and only if S ϕ's in a sufficient proportion of the relevant possible situations in which some S-trigger for ϕ-ing is present.

So much for the upshot. With both views before us, we can now focus on their relation. Things fall into place quite neatly. For as we can now see, the condition stated in SUCCESS$_{AA}$ is in fact an instance of the condition stated in SUCCESS$_{NAA}$.

Actions are successful insofar as they realize the agent's intentions. Thus, what we do by restricting the relevant possible situations to the intention situations in the case of agentive abilities is exactly what SUCCESS$_{NAA}$ postulates we do for non-agentive abilities: we restrict the relevant possible situations to the ones in which some S-trigger for ϕ-ing is present. In other words, the condition stated in SUCCESS$_{AA}$ is actually a special instance of the more general condition stated in SUCCESS$_{NAA}$. The condition stated in SUCCESS$_{NAA}$ is therefore not only the proper account of non-agentive abilities. Rather, it gives us an understanding of abilities *tout court* – of agentive and non-agentive abilities alike. The general formulation of the success view of ability is the following:

> SUCCESS$_{ABILITY}$. An agent has an ability *tout court* to ϕ (where ϕ can be an action or a mere behavior) if and only if S ϕ's in a sufficient proportion of the relevant possible situations in which some S-trigger for ϕ-ing is present.

This yields the following view. Generally, abilities are a matter of ϕ-ing in a sufficient proportion of relevant possible situations in which some S-trigger for ϕ-ing is present. In the most paradigmatic instances of agentive abilities, ϕ-ing is an action, and the crucial S-trigger situations are situations in which the agent intends to ϕ. In the case of non-agentive abilities, ϕ-ing is a mere behavior: the agent responds passively to the world. The S-trigger situations will therefore be different; they will be situations in which the agent is subjected to the world being a certain way.

The success view of agentive abilities turns out to be an instance of a broader scheme in the following way, then. Abilities are a matter of success. Success is a matter of a modal success rate. The modal success rate, in its general form, is the proportion of ϕ-ing cases among the relevant possible situations in which some S-trigger to ϕ-ing is present. When ϕ-ing is an action, the situations which matter are situations in which a suitable intention is formed.

When φ-ing is a mere behavior like hearing, digesting, or unintentionally reading street signs, the situations which matter will instead be such that the agent is subjected to the world being a certain way: to the occurrence of a sound, the ingestion of food, and the presentation of a street sign, for instance. In the case of the action as well as in the case of the mere behavior, however, the agent's φ-ing is linked to some trigger in the same normative way: φ-ing counts, in response to the trigger, as a success.

Let's consider a worry. My remarks on the relation between SUCCESS$_{AA}$ and SUCCESS$_{NAA}$ hinge crucially on the assumption that an intention is an S-trigger for an action. *A forteriori*, it hinges on the assumption that an intention is *some* trigger for an action at all. I don't find this particularly problematic. An intention, we said, is characterized by the fact that it is part of its functional role that, if nothing intervenes, it leads to action. I take it that there is a good sense in which an intention can thus be seen as a trigger for action.

Perhaps you are inclined to think that treating intentions as triggers for actions stretches the notion of a trigger beyond its reasonable boundaries. In that case, let me offer you an alternative formulation of what it is for an agent to have an ability *tout court*, which will yield the same picture I have outlined before, but without the commitment to intentions being S-triggers for actions:

SUCCESS$_{ABILITY\_}$II. An agent has an ability *tout court* to φ (where φ can be an action or a mere behavior) if and only if S φ's in a sufficient proportion of the relevant possible situations in which φ-ing counts as a success.

As is easy to see, this alternative formulation is neutral on whether or not intentions are S-triggers for actions. In the case of agentive abilities, the relevant possible situations in which φ-ing counts as a success will be intention situations. In the case of non-agentive abilities, the relevant possible situations in which φ-ing counts as a success will be situations in which some suitable external trigger is present. No commitment to actions being S-triggers for actions here. Yet, agentive and non-agentive abilities turn out to be of the same general kind, of which agentive abilities are the paradigmatic instance.

With the full view of abilities now before us, let us turn back to the stronger and the weaker reading of the success view of agentive abilities we talked about in connection with finkish abilities in chapter 4.8. On the stronger reading of SUCCESS$_{AA}$, with which I worked throughout the largest parts of chapter 4, an agent has an agentive ability to φ if and only if she φ's in a sufficient proportion of the relevant possible situations in which she intends to φ. On the weaker reading, which I presented in order to account for the finkish cases, an agent has an agentive ability if and only if she φ's in a sufficient proportion of the relevant possible situations in which she intends to ψ, where φ and ψ need not be *de dicto* or *de re* identical as long as φ-ing in response to the intention to ψ counts as a success.

Now that we have the success view of abilities *tout court* before us, these two readings can be integrated quite neatly into the overarching picture. Both the stronger and the weaker reading of SUCCESS$_{AA}$ are covered by SUCCESS$_{ABILITY}$. Thus, we get a smooth transition from non-agentive abilities where the S-trigger is an external stimulus to agentive abilities where the action type specified in the intention differs *de re* from the action type the agent has an ability for, to agentive abilities where the action type specified in the intention differs only *de dicto* from the action type the agent has an ability for, to the most paradigmatic kinds of agentive abilities, where the action type specified in the intention is identical, *de re* and *de dicto*, with the action type the agent has an ability for. In response to all of those triggers, φ-ing is a success. That is why the intention need not be the intention to φ in the case of agentive abilities – the ability is agentive simply because the S-trigger is *some* intention at all.

5.4 Abilities and dispositions

With the full picture of abilities now before us, an intriguing perspective on the relation between abilities and dispositions suggests itself. According to that perspective, abilities simply *are* dispositions of sorts – they are dispositions for successful responses to S-triggers.

When we started out thinking about abilities in chapter 1, one of the explanatory challenges for a comprehensive view of abilities we identified was to elucidate the exact relation between abilities on the one hand and dispositions on the other. That there has to be some interesting relation is obvious. The two look way too similar to be completely unrelated. Most importantly, both dispositions and abilities are modal properties in the sense that they can be instantiated without being manifested. Moreover, they both seem to be closely connected to counterfactual conditionals, both can be masked, both come in degrees, and both exhibit the corresponding kind of context sensitivity.

In view of those similarities, it is a commonplace that abilities simply *are* dispositions of sorts. In fact, the idea that abilities are dispositions is as old as the contemporary literature on abilities itself. It can be found in the writings of Ryle (1949) and Moore (1912) and has since been stated as a platitude all over the place. Ernest Sosa, for instance, simply presupposes that "abilities [...] are a special sort of dispositions, familiar examples of which are fragility and solubility" (Sosa 2011: 80).

5.4 Abilities and dispositions

Recently, the view that abilities are dispositions of sorts has enjoyed new popularity, thanks to the so-called new dispositionalists[3] (Vihvelin 2004; Fara 2008; Smith 2003), who have used it as a foundation for a new route to a compatibilist solution of the free will problem.[4] Vihvelin, for instance, writes that "[t]o have an ability is to have a disposition or a bundle of dispositions" (Vihvelin 2004: 431).[5] Fara, in much the same spirit, defends the view that "[a]n agent has the ability to φ in C if and only if she has the disposition to φ when, in C, she tries to φ" (Fara 2008: 848).[6]

I think the new dispositionalists and all the others are more or less right. Some abilities are indeed plausibly thought of as dispositions of sorts. But, and here we have made some real progress, we are now in a position to understand *why* this is and just *what* sort of dispositions they are. Consider again the success view of non-agentive abilities:

> SUCCESS$_{NAA}$. an agent S has an *non-agentive* ability to be engaged in some mere behavior φ if, and only if, S φ's in a sufficient proportion of the relevant possible situations in which some S-trigger for φ-ing is present.

To have the ability to hear, say, one has to hear in a sufficient proportion of the relevant possible situations in which one is subjected to sound, to have the ability to digest, one has to digest in a sufficient proportion of the relevant possible situations in which one has ingested food, and to have the ability to unintentionally read street signs, one has to unintentionally read street signs in a sufficient proportion of the relevant possible situations in which one is subjected to street signs. In all of those cases, one shows a successful response in a sufficient proportion of the relevant possible S-trigger situations.

---

3 The term has been coined by Clarke (2009).
4 That they have used the idea that abilities are dispositions of sorts to pursue a compatibilist strategy is what makes them *new* dispositionalists in the first place. The idea that abilities are dispositions of sorts itself is not new at all, as the other quotes above show.
5 That Vihvelin speaks about bundles is rooted in the fact that she distinguishes between basic abilities – roughly: abilities to perform basic actions – and complex abilities – roughly: abilities to perform complex actions. Basic abilities correspond directly to individual dispositions, on her view. Complex abilities correspond to bundles of dispositions (Vihvelin 2004: 431).
6 Fara's view may look a lot like the conditional analysis, but in fact, it is to be read differently. "φ-ing when trying" is not to be understood in terms of the simple counterfactual, on Fara's view. Instead, Fara wishes to remain neutral on how the modal connection is to be spelled out. Fara's view is therefore fully compatible with the success view. Indeed, the success view offers one way of spelling out what Fara suggests.

Now let's do a little experiment: let's drop the success condition. Let us postulate that one has a non-agentive ability to ϕ if, and only if, one ϕ's in a sufficient proportion of the relevant circumstances in which *some trigger* for ϕ-ing is present. No mention of the requirement that the trigger has to be such that ϕ-ing in response to that trigger counts as a success. Then it seems that we can generate counterexamples to the analysis quite easily.

Sugar dissolves in a sufficient proportion of the relevant possible situations in which it is placed in a liquid, but it seems odd, to say the least, to say that sugar has the *ability* to dissolve. Klaus Kinski used to freak out whenever someone provoked him, but it does not seem to be the proper description to say that Kinski therefore had the *ability* to freak out. Slash needed a pacemaker when he was 35, because his heart failed to pump blood in a potentially deadly proportion of all kinds of circumstances, but it would be flat-out bizarre to say that Slash's heart therefore had the *ability* to fail to pump blood. (Compare this to the perfectly fine statement that the heart of a healthy person has the ability to pump blood. – Here, the success relation is respected; hence, the ability statement makes perfect sense.)

What sugar, Klaus Kinski and Slash have are not abilities. What they have is something else: dispositions. Sugar has the disposition to dissolve, Klaus Kinski had the disposition to freak out, and Slash's heart had the highly dangerous disposition to fail to pump blood.

This suggests the following picture. First, take the following plausible view of dispositions:

> DISPOSITION. An object X has a *disposition* to ϕ if and only if X ϕ's in a sufficient proportion of the relevant possible situations in which *some* trigger – in the case of a disposition, a stimulus – for ϕ-ing is present.[7]

Let's square this with the success view of abilities:

---

[7] Note that DISPOSITION also entails an existential requirement: DISPOSITION can only be met if there is a relevant possible stimulus situation to begin with; otherwise the proportion is unspecified. This may seem wrong at first. There is a good sense in which a rubber ball is disposed to bounce when hitting the floor, even if it is nailed to the wall. (Fara (2008: 852) disputes this, but see Clarke (2009) and Whittle (2010) for a good case against Fara). And this finding seems to be at odds with the existential requirement. Note, however, that when contemplating dispositions, we usually hold only intrinsic features of the object fixed across the relevant situations. Given that only those features are held fixed, there will *always* be relevant situations in which a stimulus is present. And therefore, the sense in which the rubber ball retains the disposition to bounce is, in fact, smoothly accounted for by DISPOSITION. In section 5.7, I'll develop an analogue of this argument in more detail in connection with non-agentive abilities.

SUCCESS$_{ABILITY}$. An agent S has an *ability* to φ if and only if S φ's in a sufficient proportion of the relevant possible situations in which some *S*-trigger for φ-ing is present.

If this is true, then it is quite natural to think that abilities are simply dispositions for successful responses to S-triggers. Or, to put it differently: it seems natural to think that abilities are dispositions to φ in response to triggers, in response to which φ-ing counts as a success. The success condition is the missing piece in distinguishing abilities from dispositions, on that view. That is why it is perfectly fine to say that my heart has the ability to pump blood, whereas it is obviously bizarre to say that Slash's heart has the ability to fail to pump blood. Slash's heart has a disposition to fail, not an ability. As Millikan puts it: "No matter what dispositions a mechanism happens to have, what determines its abilities is what it was selected for doing" (2000: 63). To keep our terminology sharp, let us say that a disposition is a *mere disposition* when it is a disposition, but not an ability.

I said that this view is an intriguing one, and I am going to talk a bit more about the instructive perspective it provides in a moment. To avoid misunderstandings, however, let me add one caveat right away. The view about abilities and dispositions I have been offering depends on one crucial assumption: the assumption that DISPOSITION is the proper view of dispositions. The grounds on which this assumption seems plausible are that abilities and dispositions look so similar that there is good reason to assume that their analyses will look very similar as well. Thus, given that the success view is correct for abilities, DISPOSITION seems like a very reasonable view of dispositions.

Of course, this does not *prove* DISPOSITION to be the right view of dispositions, nor does it do anything to annihilate the arguments that have been provided for alternative accounts of dispositions (e.g. Vetter 2015; Fara 2005; Mumford 1998; Choi 2006, 2008) or against views of dispositions that have the overall structure that DISPOSITION exhibits (e.g. Vetter 2015). The objective of this book has been to come up with a viable view of abilities. It has not been to argue for or against a certain view of dispositions. Thus, I am well aware that DISPOSITION does by no means follow from the success view of abilities, and I am just as aware that it is controversial. Given that the success view of abilities is on the right track, however, DISPOSITION is *prima facie* a reasonable view to entertain as well.

What I want to argue for now is that the picture one gets by endorsing not only the success view, but also DISPOSITION, is an intriguing one. It provides a promising understanding of the similarities, but also of the differences between dispositions and abilities.

The similarities, I take it, are sufficiently explained by the fact that abilities turn out to be special instances of dispositions. Let's look at the differences, then. What is often emphasized is that abilities are active, but dispositions are passive. In the case of abilities, agents bring about change; in the case of dispositions, objects suffer change (Aristotle 1999, 9.1 & 9.8; Locke 1690, II).[8]

With the tripartite structure [dispositions – non-agentive abilities – agentive abilities] in mind, it should be clear that the idea that abilities are active and change is brought about, whereas dispositions are passive and change is suffered does not quite get things right at the outset. Non-agentive abilities, after all, seem no more active than dispositions. In both cases, change is not brought about, but suffered. Sugar dissolves when placed in liquid. I hear your voice when you shout in my ear. In terms of activity on the subject's part, these two seem rather on a par.

What is true is that *agentive* abilities are active in ways that non-agentive abilities and mere dispositions are not. And that, I take it, can be explained just fine when we look at the fact that in the case of agentive abilities, the S-trigger is the agent's very own intention. It is the agent herself who sets the trigger for change. To have an ability is for her to show the proper response to that trigger in sufficiently many cases. In this respect, agentive abilities differ from non-agentive abilities and mere dispositions. It is because of this feature that agentive abilities are active in ways that non-agentive abilities and mere dispositions are not.

The tripartite structure [mere dispositions – non-agentive abilities – agentive abilities] brings to light that non-agentive abilities have just as much in common with mere dispositions as they have in common with agentive abilities. Non-agentive and agentive abilities share the success relation between trigger and response. But non-agentive abilities and dispositions share their passiveness. In a way, then, non-agentive abilities are a middle case between the paradigmatic cases of abilities (agentive abilities) and the paradigmatic cases of dispositions (mere dispositions). I take this to be a very welcome result of the picture I am suggesting. There is something to the view that there is no sharp division between dispositions and abilities. Instead, they form a spectrum with paradigmatic cases of each at the sides and a transition area in between.

What I just said needs to be qualified a bit. It seems intriguing to think that abilities just are dispositions of sorts, full-stop. But at that level of generality, this may not be quite right. On a widely held view, dispositions are intrinsic to their

---

[8] For an interesting recent assessment of active and passive powers, see Mayr (2011: 204 ff.).

bearers.[9] The same is not true, that generality, for abilities. *General* abilities are indeed often a matter of intrinsic properties of their bearers. But specific abilities certainly aren't. Hence, I take it to be clear that specific abilities are not dispositions, if dispositions are intrinsic to their bearers.

This is an important insight. New dispositionalists (Vihvelin 2004, 2013; Fara 2008; Smith 2003) argue that the insight that abilities are dispositions of sorts offers a new and very immediate route to compatibilism about freedom and determinism. But the insights from the last paragraphs show that their argument to that effect fails.

New dispositionalists argue that since (i) abilities are dispositions, and (ii) the having of a disposition to φ is fully compatible with it being determined that the disposition remains unmanifested, (iii) the having of the ability to act otherwise is fully compatible with it being determined that one will not act otherwise.

This line of argument is faulty. It is true, of course, that dispositions can be had, even if it is determined that they will remain unmanifested. Sugar is soluble, even if it is determined that it will never be placed in liquid. But that is because solubility is a matter of an object's intrinsic properties. In determining the relevant situations, we therefore vary whether or not it is determined that the sugar will ever be placed in liquid (since we only hold the sugar's intrinsic features fixed). Thus, there will be a vast variety of relevant possible situations in which the sugar is, in fact, placed in liquid. Among those, there will be a vast proportion of cases in which the sugar dissolves. DISPOSITION is therefore met for the sugar, even if the world is deterministic. But the reason for that is that the sugar's intrinsic facts are all that matters when it comes to determining the relevant situations.

In the case of abilities, it is very different. It is uncontroversial that the ability to act otherwise has to be understood as the *specific* ability to act otherwise for it to be relevant to the free will issue. Virtually no one doubts that determinism is incompatible with agents' *general* abilities to act otherwise.[10] What is at issue has always been whether or not their specific ability to act otherwise is likewise compatible with it being determined that they will not act otherwise.

The important point is that specific abilities are abilities in view of the agent's actual situation. Thus, the ability to act otherwise, as relevant for the free will question, is the ability to act otherwise in view of all kinds of extrinsic facts about the situation. It is far from clear that that ability is had, even if it is

---

9 For counterexamples to this view, see McKitrick (2003).
10 For an exception see Keil (2007).

determined that one will not act otherwise (Clarke 2009). Whatever it is that determines that one will not act otherwise will be part of the situation one is in. Once that feature is held fixed, one will not come out as having the ability to act otherwise anymore. No one has the ability to act otherwise in view of the fact that it is determined that she will not act otherwise. That much, I take it, is clear.

Thus, the important question is whether or not the features of the situation which determine that the agent will not act otherwise do, indeed, have to be held fixed when thinking about the agent's specific ability to act otherwise. Surely, this question cannot be answered by pointing out that dispositions are retained even when it is determined that one will not manifest them. In contrast to dispositions, the abilities that matter for freedom are not intrinsic to their bearers. Since they are not intrinsic, determinism may, in fact, be a threat to the ability to act otherwise that is relevant for free will.

So much for my criticism of the new dispositionalist route to compatibilism. Let me now return to the relation between abilities and dispositions more generally. On the view I want to advance, some abilities to ϕ – the ones we have in view of our intrinsic properties – are dispositions to ϕ in response to some S-trigger. Other abilities – the ones we have in view of extrinsic properties – may not be dispositions in the ordinary sense of the word, but structurally, they resemble dispositions closely.

5.5 The normativity of abilities

I have argued for the following account of agents' abilities:

> SUCCESS$_{ABILITY}$. An agent S has an ability *tout court* to ϕ if and only if S ϕ's in a sufficient proportion of the relevant possible situations in which some S-trigger for ϕ-ing is present.

An S-trigger, I said, is a trigger in response to which ϕ-ing counts as a success. The obvious question to ask in view of my proposal is: when does a performance of ϕ-ing count as a success in response to some trigger? When ϕ-ing is an action, the answer is clear: an action is a success if and only if it realizes the agent's intentions. When ϕ-ing is a mere behavior, things are less clear. So let's try and see if we can say a little bit more about the conditions for a mere behavior to count as a success in a situation. To dispel any too-optimistic expectations right away, let me say that I will not be able to provide anything close to a comprehensive account of the success conditions of mere behaviors in this book. Despite the fact that the ideas I will present will be somewhat tentative, however, they will, as I hope, nevertheless be informative.

The most important thing to note is that, by and large, whether or not ϕ-ing counts as a success in response to some trigger depends on whether or not its occurrence in response to the trigger contributes to some end that matters in a given context. And by "context" I mean an ascriber context: whether or not we can correctly ascribe an ability to ϕ will hinge on whether or not some end is salient in our own context to which ϕ-ing contributes.

Consider a plant that absorbs toxic substances from the air and detoxifies them. Usually, this will count as a success, because it is one of our ends, as a species, to live in a non-toxic environment. Thus, we will be naturally inclined to describe the plant's potentiality (a roof term I shall use for mere dispositions and abilities alike)[11] as an ability: the plant has the ability to absorb toxic substances.

This is not to say that the end to which ϕ-ing contributes has to be the speaker's own end. Consider an evil tyrant, whose explicit end it is to develop a chemical weapon to destroy a human population in a certain area. Suppose further that our anti-toxic plant covers large parts of that area and that the plant absorbs and neutralizes the very substances that the tyrant intends to use for his chemical weapons. He may well say, in that case, "The weapon will be ineffective in that area. The area is covered by plants with the ability to absorb the toxic substance we are using".

If the speaker's own ends were crucial for whether or not the absorption of toxics is a success or not, it would surely not be a success. But that does not matter. There is *some* end towards which the absorption of toxics contributes, namely people's end to survive. As long as there is some such end, one can take the success perspective, which makes ability-talk feasible.

That is not to say that we *have to* ascribe an ability whenever there is some end to which ϕ-ing contributes. We may always just as well be neutral with respect to the success dimension of some doing. In the case of the plant, we can just as well abstain from ability-talk altogether and simply ascribe the disposition to absorb toxic substances. Whether or not an ability gets ascribed depends on whether or not the speaker chooses to emphasize the success dimension or not. But as soon as there is an end to which ϕ-ing contributes and which is salient in our context, ability talk is *feasible*.

In the case of the anti-toxic plant, the end that mattered was our end, as a human species, to live in a non-toxic environment, and – more generally – to be

---

[11] I follow Vetter's (2015) use of the notion here.

healthy and survive. Which end matters in a given context will vary, though. I have the ability to unintentionally read street signs. My heart has the ability to pump blood. My dog has the ability to smell to a much higher degree than I do. With respect to all of those abilities, it is true that there is some end, to which the respective exercises contribute. My unintentionally reading street signs contributes to my effortless information. My heart's pumping blood contributes to my survival. My dog's smelling so well contributes to his orientation and – more importantly – finding his toys behind shelves and sofas.

Underlying these are, I take it, biological functions (Millikan 1984, ch. 1 & 2). We have some system in our brains whose function it is to compute information without our conscious effort. My heart has the function of pumping blood. My dog's sense of smell has the function of giving him orientation and finding things. Functions are teleological – they have some aim built into them. So in many cases, the ends that matter for the success of some exercise will simply stem from the goal-directed nature of the biological functions had by things.

In other cases, the ends that matter will derive from other sources. The toaster has the ability, and not just the disposition, to toast bread, because toasting bread when bread is put into the toaster is what the toaster is designed to do. It is an end that derives from the intentions of the toaster's designers. Again, the *telos* is given by the toaster's function, but this time, the function is not biological in nature, but derives from someone's intentions.

The important point is not to come up with an extensive list of sources of ends, however. The important point is to see that ability-talk is feasible as soon as there is some end towards which the exercise of some ϕ contributes, where ends may stem from various sources, such as biological functions, design functions, agents' desires and intentions, and so forth.

5.6 The existential requirement revisited

When we thought about agentive abilities, it became obvious that the condition stated in the success view should be interpreted in such a way that it entails what I called "the existential requirement". To have an agentive ability to ϕ, the agent has to be able to form the intention to ϕ to begin with; this was spelled out in terms of the requirement that there have to be relevant intention situation to begin with; the set of the relevant intention situations must not be empty (→ 4.4).

The need for the existential requirement was the import from cases like that of the brainwashed follower of a cult and the coma patient. In most contexts, we obviously want to say that these agents lack the ability to raise their arm and leave the cult, respectively. And the reason for that is that the existential requirement is violated in their cases. Holding their respective conditions fixed, there is no relevant intention situation to begin with; they cannot leave the cult and raise their arm, respectively, because they cannot form the relevant intentions to do so in the first place.

The existential requirement could have been formulated as a second condition for what it is for an agent to have an ability. Instead, I showed that the original condition – the requirement that the agent φ in a sufficient proportion of the relevant intention situations – already entails it. For the proportion of φ-ing cases among the relevant intention situations to be sufficiently high, there have to be relevant intention situations to begin with. If not, the proportion is an unspecified ratio (because we are dividing by zero). Thus, the existential requirement is entailed as a necessary condition by the success view of agentive abilities.

In this chapter, I have extended the success view to cover not only agentive, but also non-agentive abilities. I said that having an ability, quite generally, is a matter of φ-ing in a sufficient proportion of relevant S-trigger situations. The success view for agentive abilities turned out to be an instance of this more general scheme. In the case of actions, the S-trigger we look out for is always the agent's intention.

The analysis of non-agentive abilities (and abilities *tout court*) I developed in this chapter is structurally analogous to the success view of agentive abilities. Thus, it, too, entails an existential requirement. For it to be the case that there is a sufficient proportion of φ-ing cases among the relevant possible S-trigger situations, there have to be relevant possible S-trigger situations to begin with. This, then, is the more general version of the existential requirement.

It may not be entirely obvious that this is the right verdict in the case of non-agentive abilities. In the remainder of this section, however, I will argue that it is in fact exactly right. Non-agentive abilities, too, are had only if the set of relevant S-trigger situations is non-empty.

When we thought about agentive abilities in the last chapter, we saw that there are two structural grounds on which an agent can lack an ability. The specific ability to raise one's arm, for instance, can be impeded in two different ways. On the one hand, the agent may wear a strait jacket. On the other hand, the agent may be in a coma.[12]

---

[12] If the agent is permanently comatose, the case is better described as one in which the agent also lacks the general ability. We are now concerned with a different case: a case in which the agent has the general ability, but lacks the specific one. I take it that the agent undergoing a temporary and brief state of coma is one way in which this constellation can come about.

In the first case, the agent lacks the ability, because the agent's modal success rate is too low. The ratio expressing the proportion of performance cases among the relevant intention situations is specified, but it is zero. In any context, that will be too low for us to be willing to count the agent as having the ability.

In the second case, the agent lacks the ability for other reasons. Here, the ratio corresponding to the proportion of performance cases among the relevant intention situations is not specified to begin with. The agent cannot intend to raise her arm – thus, the relevant possible situations do not contain any intention situations. The agent lacks the ability to raise her arm because the existential requirement fails to be met.

In the case of non-agentive abilities, the same two grounds for an agent lacking the ability can be found. On the one hand, the proportion may simply be too low; on the other hand, the relevant possible situations may not contain any S-trigger situations to begin with. Take the non-agentive ability to unintentionally read advertisements, and let's focus on the specific ability to do so. What can be grounds for you to lack that ability? On the one hand, of course, you may be blindfolded, blinded by the sun, distracted, drugged, or otherwise impaired in such a way that you fail to read the ads in sight. All of those impediments are such that your modal success rate in reading advertisements will turn out too low.

On the other hand, however, you may lack the specific ability to unintentionally read an ad simply because there is no ad in sight. In that case, there will not be any S-trigger situations among the relevant possible situations. Hence, you lack the specific ability to read advertisements because the existential requirement fails to be met.

What this shows is that the existential requirement is important not only for agentive abilities, but also for abilities of the non-agentive kind. No one can hear in view of the fact that there is nothing to be heard, digest in view of the fact that there is nothing to be digested, and adjust their pupil size to the lighting conditions, if the lighting conditions don't change.

In the case of specific non-agentive abilities, this is quite obvious, I guess. In the case of general non-agentive abilities, a little more reflection may be needed. Take the general ability to hear. The existential requirement postulates that an agent has the general ability only if there is a relevant possible situation in which sound occurs. At first sight, this may not strike you as entirely plausible. Does not an agent retain her general ability to hear even if no sound can occur? Suppose the world is such that the atmosphere absorbs every sound signal right away – no sound can ever reach a human ear. Don't agents retain their general ability to hear in such a world?

The answer depends on *how* general the ability is we are thinking about here. There is a good sense in which agents lack the fairly general ability to hear in a world in which no sound can occur – there is just nothing to be heard, after all. But there is also a good sense in which normal agents can still hear – they still have what it takes to hear sound if it were in fact occurring. The difference between those two uses of "the ability to hear" is not that only one is a general ability; even in the first sense, we are interested in what the agents can do across a broad range of situations and not in a particular situation. We are dealing with two abilities on the general side of the spectrum. Yet, the two differ with respect to just *how* general they are.

In the first case, we are still holding a very stable feature of the world fixed: that the world is such that no sound ever occurs. In that case, the existential requirement will fail to be met. There are no relevant sound situations. Thus, the agent lacks the ability to hear. This seems right. In view of the fact that the agent lives in a world with no sound, the agent cannot hear sound.

In the second case, we are holding only the agent's intrinsic hearing related make-up fixed. That the world is such that no sound ever occurs is varied across the relevant possible situations, then. Thus, there *will* be relevant sound situations. And among those, there will of course be a vast number of situations in which the agent hears. Again, this is correct. In view of her hearing organs, the agent can hear.

To my mind, the second use seems like the more natural one. The ability to hear, understood as a general ability, is usually given the fully general interpretation where only the agent's stable intrinsic abilities matter. In that case, the existential requirement will *always* be met. Since we vary what the world is like, there will always be S-trigger situations among the relevant possible situations.

5.7 Proportions among infinite sets

There is an obvious difficulty with the success view as it stands. It is not clear how to make sense of the notion of proportions of situations, as it figures in the account. That is because many sets of situations will have infinite cardinality. And the proportion of the infinite set among the infinite set will always be 1.

Take an example. Consider my ability to jump in view of my muscular constitution. According to the success view, I have that ability if and only if jump in a sufficient proportion of the relevant possible situations in which I intend to jump. The relevant possible situations will be those in which my muscular constitution is held fixed. That set of situations will presumably be infinite. The length of each of my hairs will vary across the relevant possible situations, and since the lengths of my hairs are continuous, there will be continuously many situations in which my muscular constitution is held fixed.

The same goes for any subset of the relevant possible situations, which allows for variances along the continuous scale. The set of the relevant possible situations in which I intend to jump are likewise infinite, because it will still contain as many situations as there are lengths of my hairs. And likewise for the relevant intention situations in which I jump. They, too, constitute an infinite set.

If that is true, the success view faces a fundamental problem. It seems to lack the resources to distinguish between cases in which an agent has an ability and cases in which an agent lacks an ability. The proportion of ɸ-ing cases among the relevant intention situations will pretty much always be v, such that v = the size of the infinite set/the size of the infinite set. Obviously, this is a serious worry.

How can the problem be mitigated? First, it is important to note that the challenge we are facing is not in any way peculiar to the success view. As we are going to see in the next, and last chapter, every contemporary account of abilities is in some way or other committed to the notion of proportions of situations. Greco (2007) analyzes abilities in terms of the requirement that the agent has to be successful in a suitable proportion of a certain set of worlds (where "worlds" should read "situations"). Maier (2013) analyzes general abilities in terms of a suitable proportion of similar circumstances. And Sosa (2015) argues that

> archery competence, for example, is a sufficient spread of possible shots (covering enough of the relevant shapes and situations one might be in as an agent) where one would succeed if one tried, an extensive enough range. (Sosa 2015: 98)

The same goes for recent views of dispositions. Both Manley and Wasserman (2008) and Vetter (2015), for instance, talk about proportions of (centered) worlds in their accounts of dispositions. What this shows is that abilities (and dispositions) are generally taken to have *something* to do with proportions of situations. Once we agree on that, we are in the same boat when it comes to problems concerning the comparability of sets of such circumstances.

Apart from the fact that it is hard to think of a way to avoid proportions of worlds or situations in an account of abilities, I take it to be clear that we should also have a more general interest in a formal model which allows for there being more circumstances of one sort than of another. There is, for instance, good reason to think that *graded possibilities* have something to do with proportions of worlds, and not, as Kratzer (1981) prominently holds, with closeness to an ordering source. Likewise, and perhaps relatedly, there is good reason to think that *probabilities* have something to do with distributions of events across worlds – the probabilities for an event are higher, on that account, the larger the set of (relevant) worlds in which the event occurs (Bigelow 1976).

That both graded modalities as well as probability seem to call for some measure for sizes of sets of worlds is instructive to note. Graded possibility, probabilities, and abilities seem very closely connected, after all. Suppose p is more easily possible than q. Then one should think that the probability of p is higher, according to some measure of probability, than that of q. Likewise the other way around. If p is more probable than q, then one should think that p is more easily possible, given some measure of possibility, than q. Abilities seem closely connected to both. If an agent S1 is more able to ϕ than another agent S2, then one should think that it is more easily possible, according to some measure of possibility, and more probable, according to some measure of probability, that S1 ϕ's upon intending to ϕ than that S2 ϕ's upon intending to ϕ.

If probabilities, graded modality, and abilities turned out to be a matter of proportions of worlds, things would fall into place quite neatly. If, on the other hand, talk about proportions of worlds turned out to be insensible after thorough investigation, then my suggestion would be to just work with a probability framework instead.

I don't think too much pessimism is called for, though. Just like Manley and Wasserman (2008), I take it that there is good reason to think that proportions of worlds or situations can in fact be made sense of. The key lies in measure theory. As Bigelow (1976) has argued in an attempt to define probability in terms of similarity between worlds, we can project a similarity measure onto modal space which yields distances between worlds. Once we have those distances, we can measure the region that is covered by any set of possible worlds.

If that is true, then the proportion of ϕ-ing cases among the relevant intention situations, which is needed to make sense of the success view as it stands, can be interpreted as v, such that v = the size of the region of the relevant intention cases in which the agent ϕ's/the size of the total region of relevant intention situations.

The challenge, of course, is to come up with a similarity measure that yields the right distances for some given purpose. This task is far from trivial, and it would require another book to show how Bigelow's approach can be put to work in connection to the success view. In principle, however, measure theory provides the formal equipment to tackle this task.

5.8 A brief reflection on the explanatory challenges

Before closing this chapter, let me briefly get back to the explanatory challenges that we identified for a comprehensive view of abilities in chapter 1. In the previous chapters, answers to those challenges have emerged, but I we have not, so far, explicitly mapped them onto the challenges. Let me therefore go through the challenges one by one and briefly say a few words about each before moving on with the discussion of the success view.

The explanatory challenges, recall, were the following:

E1: From a comprehensive view of abilities, we may hope to learn how abilities relate to counterfactuals.

E2. From a comprehensive view of abilities, we may hope to learn about the relation between abilities and dispositions.

E3. From a comprehensive view of abilities, we may hope to learn how ability ascriptions relate to the semantics for "can" ascriptions in general.

With respect to E1, it has become evident that the relation between abilities and counterfactuals is by no means straightforward. As I said in chapter 1.5, there is certainly *some* connection between abilities and counterfactuals – if someone has an ability, it will often be true that the agent would exercise that ability, if she intended to –, but, as we know by now, that connection is not strong enough to allow for an analysis of abilities in terms of a counterfactual conditional.

That is not to say that an analysis of abilities in terms of counterfactuals in general is doomed. In fact, we will see in chapter 6.2 that there is a way of formulating the success view itself in counterfactual terms. What we will see too, however, is that the counterfactuals that play a role in such an account bear only very little resemblance to the counterfactual that figured in the simple conditional analysis and which seems intuitively linked to abilities. In fact, they bear so little resemblance to counterfactuals that the view had better be expressed differently. Quite generally, the upshot will be that counterfactuals do not play any central role in an account of abilities.

With respect to the explanatory challenge of providing information about the relationship between abilities and dispositions, we can be very brief. On that point, we reached the verdict that abilities and dispositions are closely related indeed, in that (intrinsic) abilities just seem to *be* dispositions of sorts; roughly: dispositions for success. This verdict hinges on the plausibility of the view of dispositions I put forward in section 5.6, of course. But insofar as that view is on the right track, (intrinsic) abilities turn out to be a special kind of dispositions.

Finally, let me say a few words about the relationship of ability ascriptions and "can" statements quite generally. As laid out in chapter 1.7, the challenge here is to either align one's views of abilities with the standard semantics of "can" statements as championed by Kratzer (1977, 1981) or reject the standard semantics when it comes to ability statements and thereby reject a core feature of that semantics, namely the claim that "can" statements exhibit a unified structure in that all of them express possibility.

It should be clear that my verdict on that matter is the latter. *Pace* the standard semantics, "can" statements which ascribe ability are not possibility statements (→ 2). Hence, it is not true that all "can" statements express possibility. But of course, some "can" statements do express possibility. And hence, "can" statements do not exhibit a unified structure. The standard semantics goes astray not only in its assessment of ability statements, but also in its assumption that "can" statements form a perfectly unified class semantically.

5.9 Upshot

Let's wrap up. Abilities, I argued, are a matter of success. In the case of agentive abilities, this basic idea was spelled out in terms of what I then called the agent's modal success rate: the proportion of the relevant intention situations in which the agent performs an action to the relevant intention situations as a whole. In section 5.1, I argued that this account cannot be carried over to unintentional abilities one-to-one.

What *does* carry over one-to-one, however, are the structural features of the account. As I laid out in section 5.2, non-agentive abilities, too, are a matter of success. The difference is just that success is constituted differently. In the case of actions, success has to do with a sufficiently strong modal tie between intentions and performances. This is why the relevant possible situations are restricted to the ones in which the agent forms a proper intention. In the case of mere behaviors, the agent's intention does not play a crucial role. Instead, in such cases, success consists in showing the right response to certain external triggers in the world.

These insights led up to the success view of non-agentive abilities, according to which an agent S has a *non-agentive* ability to be engaged in some mere behavior ϕ if and only if S ϕ's in a sufficient proportion of the relevant possible situations in which some S-trigger for ϕ-ing is present, where an S-trigger for ϕ-ing is any trigger such that ϕ-ing in response to that trigger counts as a success.

In section 5.3, I laid out that the success view of non-agentive abilities works analogously to the success view of agentive abilities in all crucial respects. As before, the domain of the relevant possible situations will vary across contexts, which accounts for the distinction between general and specific non-agentive abilities. Moreover, the degree of an agent's non-agentive ability to φ will be determined in the same way, structurally, as the degree of an agent's agentive ability to φ. It corresponds to the weighted proportion of the performance cases among the relevant possible S-trigger situations. No big differences here.

Section 5.4 was all about integrating the success view of agentive and the success view of non-agentive abilities into a common view. Doing so was easy. As I showed, the condition figuring in the success view of non-agentive abilities actually encompasses the condition figuring in the success view on agentive abilities. Having an ability to φ, quite generally, is for the agent to φ in a sufficient proportion of the relevant possible S-trigger situations, where this boils down to φ-ing in a sufficient proportion of the relevant possible intention situations in the case of agentive abilities. Thus, we can formulate the fully comprehensive view of abilities tout court as the view that

SUCCESS$_{ABILITY}$. an agent S has an ability to φ if and only if S φ's in a sufficiently high weighted proportion of the relevant possible S-trigger situations.

In section 5.5, I laid out an intriguing perspective on the relationship between abilities and dispositions. A *prima facie* plausible take on dispositions, against the background of the success view of abilities, is that having a disposition to φ is for an object to φ in a sufficient proportion of the relevant possible situations in which a trigger for φ-ing is present. If this is right, then abilities are dispositions of sorts, namely dispositions in which the trigger that is required is an S-trigger – a trigger in response to which φ-ing counts as a success. The difference between abilities and mere dispositions is a normative difference and one of perspective, on that account.

Section 5.6 turned to this normative dimension of abilities and asked what it is for φ-ing to count as a success in response to some trigger. The answer, I suggested, is that, by and large, whether or not φ-ing counts as a success in response to some trigger depends on whether or not its occurrence in response to the trigger contributes to some end that matters in a given context. As soon as there is *some* end to which φ-ing contributes and which is salient in our context, ability talk is *feasible*.

In section 5.7, I spotlighted an implication of the success view of abilities *tout court*, namely that the existential requirement holds, not only in the case of agentive, but also in the case of non-agentive abilities. This may seem odd at first. The ability to hear, one may think, is retained, even if there is no relevant possible situation in which an S-trigger for hearing – the occurrence of sound, say – is present. Looking more closely, though, this is false.

There is a good sense in which agents lack the ability to hear when no sound can occur – there is just nothing to be heard, after all. This is the sense that is obtained in virtue of the existential requirement. Usually, though, we are interested in the ability to hear *in view of the agent's intrinsic make-up alone,* and that ability is retained when no sound can occur. That, however, does not conflict with the truth of the existential requirement. That is because in the case of the ability to hear in view of one's intrinsic features extrinsic facts alone – including the fact that no sound can occur – are going to be varied. And hence, there will always be sound-situations among the relevant possible situations in such a case. The prevalence of the sense in which one can hear even though no sound can occur is therefore unproblematic for my commitment to the existential requirement.

In section 5.8, we briefly reflected on the explanatory challenges and said that the relationship between abilities and counterfactuals is not as close as one would have thought, that abilities are plausibly understood as dispositions for success, and that the standard semantics of "can" statements goes wrong when it comes to "can" statements that assign ability.

6 The success view situated

I have developed the success view against the background of two views: the simple conditional analysis on the one hand and possibilism on the other. There was good reason for this way of setting the stage. Those two views really *are* the background for the contemporary debate on abilities. Virtually every contemporary account of abilities draws on resources that are provided by one or both of these two background views, and the struggle to overcome the shortcomings of the background views shines through every single one of them.

Yet, there are quite a few more recent views on the market, and those views, of course, need to be taken into account as well. This chapter turns to this task. My aim, in what follows, is to situate the success view among some of the recent views on abilities, point out similarities and differences, and provide reasons for favoring the success view over the other accounts.

Most of the recent interest in abilities stems from two debates that are taking place in different areas of philosophy. On the one hand, there is the so-called "new dispositionalism" (Vihvelin 2004, 2013; Smith 2003; Fara 2008). As noted before (→ 5.4), new dispositionalists think that abilities are dispositions of sorts. Of course, that idea is not very new (Ryle 1949; Moore 1912). What makes new dispositionalism *new* is the combination of that idea with two further claims.

First, new dispositionalists share the conviction that understanding the link between abilities and dispositions provides an immediate and new route to compatibilism about freedom and determinism. In chapter 5.4, I argued that the line of reasoning leading up to this conviction is inconclusive.

The second conviction characteristic of the new dispositionalism is more important for the points that will concern us in the course of this chapter: new dispositionalists unanimously reject the simple conditional analysis of dispositions. According to the simple conditional analysis of dispositions (Moore 1912; Ryle 1949), an object has a disposition to M when C (to dissolve when placed in liquid, say) if and only if the object would M, if C were the case (if it would dissolve, if it were placed in liquid).

Just as in the case of abilities, it is nowadays widely believed that dispositions cannot be analyzed in terms of a single counterfactual. In part, the reasons for the rejection of this view are analogous to problems that have also played a role in our discussion of the simple conditional analysis of abilities. Just like abilities, dispositions can be masked, come in degrees, and ascriptions of dispositions are correspondingly context sensitive. My ability to fall asleep at night can be masked by too much street noise (Fara 2005: 50); one thing can be more fragile than another (Manley & Wasserman 2008: 72); and what counts as fragile in a discussion among engineers need not count as breakable in the context of a discussion among ceramists (ibid.).

What we can note, then, is that a single counterfactual is just as ill-suited when it comes to accounting for dispositions as it is when it comes to accounting for abilities. New dispositionalists are well aware of this fact. That is why they reject the simple conditional analysis.

Which view of dispositions do they endorse instead? Here, new dispositionalists diverge. The new dispositionalists are not primarily interested in a positive account of dispositions. Their primary concern is to provide an account of abilities *in terms of dispositions*. Only sometimes is that account then supplemented by a positive account of dispositions.

Thus, new dispositionalist views about abilities come in different granularities, if you so will. On the coarse-grained level, new dispositionalists provide an analysis of abilities *in terms of dispositions*. Thus, Fara suggests that "[a]n agent has the ability to ϕ in C if and only if she has the disposition to ϕ when, in C, she tries to ϕ" (Fara 2008: 848).[1] And Vihvelin writes that "[t]o have an ability is to have a disposition or a bundle of dispositions" (Vihvelin 2004: 431).[2] On the coarse-grained level, then, new dispositionalists are more or less *d'accord* on how to analyze abilities.[3]

On a more fine-grained level, this need not be the case. New dispositionalists need not, and do not, agree on how to analyze dispositions. Thus, new dispositionalist views on abilities can, and do, differ once various different views on dispositions are plugged into the coarse-grained analysis. On a more fine-grained level, then, new dispositionalist views on abilities can, and do, take rather different forms, depending on the account of dispositions which is endorsed over and above the coarse-grained analysis of abilities in terms of dispositions.

How does the success view relate to the coarse-grained and various existing and potential fine-grained versions of the new dispositionalism? On the face of it, my views are fully in line with the coarse-grained new dispositionalist view

---

1 Fara's view may look a lot like the conditional analysis, but is to be read differently. See fn. 62.
2 That Vihvelin speaks about bundles is rooted in the fact that she distinguishes between basic abilities – roughly: abilities to perform basic actions – and complex abilities – roughly, abilities to perform complex actions. Basic abilities correspond directly to individual dispositions, on her view. Complex abilities correspond to bundles of dispositions (2004: 431).
3 Vihvelin (personal conversation) thinks that only what I have called agentive abilities can be analyzed in terms of the disposition to ϕ when trying to ϕ. Non-agentive abilities correspond to other dispositions, on her view.

that abilities are dispositions of sorts. As I laid out in chapter 5.6, I find it very plausible that

> DISPOSITION. an object has a disposition to φ if and only if the object φ's in a sufficient proportion of the relevant trigger situations for φ-ing.

Combine this plausible analysis with my view that having an ability to φ is for an agent to φ in a sufficient proportion of a subset of the relevant possible S-trigger situations and what you get is the new dispositionalist's credo that abilities simply are special instances of dispositions.

Note, however, that I am not *analyzing* abilities in these terms. I am not *committing* myself to the view that abilities are dispositions. Instead, I am committing myself to a certain analysis of abilities – the success view – and to the further hypothetical statement that *if* dispositions are analyzable along the lines of DISPOSITION – which I take to be plausible, but by no means certain – *then* abilities turn out to be dispositions of sorts. Thus, I may be *in line* with the new dispositionalist credo that abilities are dispositions, but I am in no way committed to it.

The coarse-grained new dispositionalist view of abilities will not concern us any further in the present chapter. We will, however, be concerned with a few analyses of abilities that emerge when one plugs different analyses of dispositions into the coarse-grained new dispositionalist picture that abilities are dispositions of sorts. More specifically, we will look the sophisticated conditional analysis which is obtained by analyzing abilities in terms of Manley and Wasserman's (2008) very powerful account of dispositions (→ 6.2) and the view one obtains by analyzing abilities in terms of Vetter's (2015) extended possibilist account of dispositions (→ 6.3).

Let me now proceed to talk about the other strand of literature in which abilities have played an important role quite recently: the literature emerging from the so-called virtue epistemology movement.

"Virtue epistemology" is an umbrella term for a rather wide variety of views, which differ quite substantially in many ways. According to one strand of those views, abilities have to be considered to be at the heart of epistemic properties, such as knowledge or justification. Among the virtue epistemologists who phrase their views in terms of abilities, there have been serious attempts to understand that notion better. Let me give you a little bit of background on that.

As diverse as virtue epistemological accounts are on many levels, there are two basic commitments that unify all of them (Greco & Turri 2015). The first is that virtue epistemologists consider epistemology to be a normative discipline.

In Greco and Turri's words, that "implies that epistemologists should focus their efforts on understanding epistemic norms, value and evaluation" (2015).

Different virtue epistemological accounts spell this out differently. While all virtue epistemologists agree that epistemic norms, value, and evaluation play a crucial role in understanding epistemological concepts, there is a divide between virtue epistemologists who believe that those concepts cannot be analyzed in purely non-normative terms (e.g. Zagzebski 1996, 2009) and those who think that they can (e.g. Greco 1999, 2009; Sosa 2007).

The second commitment common to all virtue epistemological views is that "intellectual agents and communities are the primary source of epistemic value and the primary focus of epistemic evaluation" (Greco & Turri 2015). Greco and Turri explain the general line of thought by way of a very helpful example:

> An evidentialist might define an epistemically justified belief as one that is supported by the evidence, and then define evidence in a way that entirely abstracts from the properties of the person. On such an approach, it would be natural to understand intellectual virtues as dispositions to believe in accordance with the evidence (which, again, is defined independently, without mentioning the virtues). A virtue epistemologist would reverse the order of analysis, defining justified belief as one that manifests intellectual virtue, and evidence in terms of intellectual virtue. (2015)

With this very brief introduction into the two main ideas of virtue epistemology in the background, let us see how abilities enter the picture.

Abilities play a central role in one very prominent strand of virtue epistemology which stands in the tradition of reliabilist accounts of knowledge. On those virtue epistemological views, knowledge is non-accidentally true belief, where this, in turn, is spelled out in terms of belief that is true because it has been obtained in an intellectually virtuous way and not by mere coincidence (Sosa 1991; Greco 2009).

What is it for a belief to turn out right by reason of virtue and not just by coincidence? Here, many virtue epistemologists have spelled out their view by using the notion of an ability. Greco, for instance, lays out that

> [t]o say that someone knows is to say that his believing the truth can be credited to him. It is to say that the person got things right due to his own abilities, efforts and actions, rather than due to dumb luck, or blind chance, or something else. (Greco 2003: 111)

That a belief is obtained via one's intellectual or cognitive abilities is what guarantees, according to the above-mentioned virtue epistemologists, that the belief is non-accidental. That is because an ability to ϕ is had only if the agent's success in ϕ-ing is reliable enough. This is where the reliabilist intuition underlying these virtue epistemological views becomes visible. According to reliabilists (e. g. Goldman 1979), one knows that p only if one's belief that p is obtained through a reliable process. The virtue epistemological twist of this intuition is the postulate that one's beliefs be formed in virtue of one's intellectual abilities.

It should now be clear why it is quite important to supplement one's virtue epistemological view that knowledge is true belief obtained through ability with an actual account of what it is to have an ability. In order to show that abilities have to do with reliability, one obviously needs to talk about abilities quite a bit.

It is therefore unsurprising that two very promising accounts of abilities have come up in the virtue epistemological sphere. One is put forward by Greco (2007), the other one by Sosa (2015). Despite their shared virtue epistemological background, however, the views are not equivalent. Greco's view is neutral on a question that is addressed by Sosa. And while Greco's view is fully compatible with the success view, it takes a bit more effort to see how exactly the success view relates to Sosa's view of abilities: Sosa's view happens to be the very same view one obtains when plugging Manley and Wasserman's view of dispositions into the new dispositionalist credo that abilities just are dispositions of sorts. Despite the fact that the new dispositionalism and virtue epistemological accounts differ very substantially in their objectives, there are actually overlaps in the views of abilities that have emerged against these two backgrounds.

In what follows, I will not systematize the views of abilities that are going to be discussed along the new dispositionalism/virtue epistemology divide. And in fact, not all of the views I will discuss have emerged from one of those two background debates. Thus, I will focus on the views themselves and systematize them in accordance with their relation to the success view.

We will begin, in the next section, 6.1, with Greco's analysis of abilities in terms of success, which is neutral regarding some crucial details and thus compatible with a variety of other views, including the success view itself. In fact, I fully agree with Greco's view of abilities and see the success view as the proper means to spell out the account more fully.

In section 6.2, we'll move on to what I will call "the sophisticated conditional analysis". The sophisticated conditional analysis is modeled on Manley and Wasserman's (2008) account of dispositions. Moreover, it is defended by Sosa (2015), who is a virtue epistemologist and, in a more complex version, by Vihvelin (2013), who is a new dispositionalist. Enriched with some ameliorations, I will argue, the view has the same merits as the success view. But the reason for that is that the sophisticated conditional analysis *cum* ameliorations simply *is* a version of the success view. Since it is a less elegant and somewhat misleading one, I suggest we should nevertheless endorse the success view instead. Yet, the views are really very similar, and I view its proponents as allies in most crucial respects.

In section 6.3, we'll look at a view that has not actually been defended by anyone but is obtained by analyzing abilities along the lines of Vetter's (2015) account of dispositions. Since the resulting view is quite interesting from a new dispositionalist perspective, but also, and especially, in relation to the success view, I take it to be a fruitful endeavor to take a look at it in some detail and see why it will not do the job.

In section 6.4, finally, we'll look at a view which has recently been put forward by Maier. Maier's view has certainly been developed with an eye towards free will issues, but in contrast to new dispositionalist views of abilities, it does not emphasize any ideas about the relationship between abilities and dispositions. Instead, Maier thinks that abilities should be analyzed in terms of the choice theoretic notion of an option. Doing so strikes me as unfortunate in a number of ways. Accounting for abilities in terms of options, it seems to me, really gets some things wrong that can be accommodated for very smoothly by the success view.

Despite the fact that I substantially disagree with some of the views I'll discuss and think we should favor the success view over all of them, let me say one thing right away. In principle, I consider all of the views I will be discussing in this chapter as attempts to voice most of the core insights that I have been trying to voice in the form of the success view. The recent views on abilities do not differ vastly from the success view, or, for that matter, from one another. All of them, for instance, incorporate the idea that having an ability is a matter of performing (or, in the case of Maier's view, having options) across a certain proportion of worlds or situations. Regarding very many fundamental aspects, there is considerable agreement between proponents of the various views I will talk about. The devil is in the details. Let's see if we can pin it down.

6.1 Success 1.0

Let's start with Greco's view on abilities as developed in connection with his virtue epistemological endeavor to analyze knowledge in terms of agents' intellectual abilities. On Greco's view,

> SUCCESS_1.0. S has an ability A(R/C) relative to environment E = Across the set of relevantly close worlds W where S is in C and in E, S has a high rate of success in achieving R. (Greco 2007: 61)

I choose the label SUCCESS_1.0, because Greco's view is not fully compatible with the success view; rather, it voices the same basic intuition, albeit without filling in the details. The success view, as I have presented it, is a 2.0-version of Greco's account, so to speak.

To see that, let's look at SUCCESS_1.0 in some more detail. First, note that Greco is concerned with abilities relative to environments. Thus,

> Jeter has an ability to hit baseballs relative to typical baseball environments. He has no such ability relative to active war zones ... where he lacks the concentration needed to hit baseballs (ibid.).

Moreover, Greco points out, "abilities are tied to a set of relevant conditions" (ibid.). Thus, "it does not count against Jeter's having the ability [to hit baseballs] ... that he can't hit baseballs in the dark, or with sand in his eyes" (ibid.: 60).

I have to say that it is not quite clear to me how environments and conditions are supposed to differ. Greco emphasizes that there is considerable overlap between the two. The tentative difference he has in mind is just that environments are supposed to be "relatively stable circumstances", whereas conditions are "sets of shifting circumstances within an environment" (ibid.). Let's grant him that the distinction can be made sense of. Nothing hinges on this in what follows.

It should be obvious that Greco's emphasis on conditions and environments sits well with the success view. The success view postulates that an agent has an ability if and only if she ϕ's in a sufficient proportion of the relevant possible situations in which some S-trigger for ϕ-ing is present. The crucial notion here, of course, is that of the relevant possible situations. The relevant possible situations can be selected as to include only those in which the agent is in a typical baseball environment in which the agent does not have sand in her eyes, for instance. Of course, these situations will have to be otherwise similar enough to the actual world in many respects. We always hold certain features of the agent and the world fixed across the relevant possible situations. Greco distributes these two onto two distinct elements of his analysis. He talks about abilities relative to environments and analyzes those abilities in terms of relevantly close worlds.

I take it that which way one goes here is by and large a matter of taste. Let me emphasize, however, that it is often important to drop the restriction to a certain environment and conditions when thinking about abilities. That is because while it may not count against an agent's ability *simpliciter* in some context that the agent is unable to exercise it in extremely challenging circumstances, it may well count against a certain degree of her ability. If Jeter hits baseballs even with sand in his eyes or in active war zones (or both!), while his buddy Jeff doesn't, then this counts in favor of Jeter being better at hitting baseballs than Jeff. By restricting one's focus on abilities relative to environments and conditions right from the outset, one is in danger of losing sight of the fact that for some purposes it is useful to drop the restrictions to environments in order to compare success rates *across* environments. This is a minor point, though. By and large, SUCCESS_1.0 is on the right track.

Secondly, note that Greco uses "rate of success" in the same way I have used it throughout the book:[4] as a modal notion. What matters is success *across possible worlds*:

> [To say that someone has a high rate of success]is to say both more and less than that they have a good track record with respect to achieving that result. ... Actual track records can be the result of good luck rather than ability. Likewise, actual track records can be the result of *bad* luck rather than *lack* of ability. (ibid.: 60 f.)

Note that Greco speaks of possible worlds instead of possible situations. I take this to be merely sloppy and not a matter of dispute, however. Once we start looking at the rate of success, we need to look at the rate of success even within worlds. Alternatively, we will have to use "world" to mean "centered world" or the like. I don't think Greco would dispute that. Again, SUCCESS_1.0 is in line with the success view of ability.

We can note, then, that everything Greco says (or can be assumed to mean by what he says) is fully compatible with the success view as I have been advancing it in this book. But – and this is why I have been referring to it as the 1.0 version of the success view itself – it is less informative in one crucial respect. Greco postulates, roughly, that to have an ability, the agent has to be successful across sufficient range of situations (or worlds, as Greco puts it). What Greco is silent about, however, is what exactly this amounts to.

The answer to this question is by no means obvious. Greco's view is, for instance, compatible with a view according to which having an ability is a matter of φ-ing in a sufficient proportion of the relevant worlds, full-stop. No restriction to intention cases here. What we are evaluating, on this view, is just the proportion of the φ-ing situations among the totality of the situations in which the relevant facts obtain. Call this view "possibilism+". It is obtained by analyzing abilities along the lines of Vetter's (2015) account of dispositions. Vetter does not defend this view, because she sees that it results in problems, and I will walk

---

4 In fact, I have borrowed the term from him.

you to these problems in section 6.3. What is important for our concerns here is just that possibilism+ is compatible with Greco's SUCCESS_1.0.

Greco's view is also compatible with a view, according to which having an ability is a matter of a multitude of counterfactuals being true, each of which has the form "If some S-trigger for φ-ing were present and C were the case then S would φ", where C picks out some more or less fully specified situation. Call this view "the sophisticated conditional analysis". Versions of it are defended by Sosa (2015) and Vihvelin (2013), and an analogue of the view has been defended by Manley and Wasserman (2008) in the case of dispositions.

There are many ways to put flesh to the bones of Greco's account, then, and the success view offers one of those ways. Just like possibilism+ and sophisticated conditional analyses, then, the success view can be seen as a more elaborate account that is fully in line with Greco's basic view, but by no means entailed by it.

6.2 The sophisticated conditional analysis

Having identified Greco's view as fully compatible with the success view, let's move on to a view which emerges when analyzing abilities along the lines of a very powerful account of dispositions that has recently been proposed by Manley and Wasserman (2008). On Manley and Wasserman's view, which I will refer to as "the sophisticated conditional analysis of dispositions",

> SOPH_D. N is disposed to M when C if and only if N would M in some suitable proportion of C-cases. (ibid.:76)

SOPH_D is quite interesting for our concerns, because it owes its potency to the fact that it solves all of the most pressing problems of the simple conditional analysis of dispositions. The view accounts smoothly for masks, degrees, and context sensitivity. Since these are all issues that will have to be accounted for in the case of abilities as well, there is good reason to think that an analogue of SOPH_D will also prove powerful in the realm of abilities.

In what follows, I'll argue that this is indeed so. But only after adding a few crucial supplements to the sophisticated conditional analysis, and only because the sophisticated-conditional-analysis-cum-supplements will turn out to be nothing other than a less elegant and somewhat misleading version of the success view itself. It should therefore not come as a surprise that the two views are eventually on a par when it comes to their merits.

To get a better grip on SOPH_D, let's look at the view in some more detail. First, let's try and understand what C-cases are. C-cases, according to Manley and Wasserman, are first of all cases in which a proper stimulus is present. For solubility, C-cases are those in which N is placed in a liquid; for fragility, C-cases are cases in which N is subjected to force; for irascibility, C-cases are cases in which N is provoked. Some dispositions have manifold stimuli; in that case, the C-cases comprise cases in which any of those stimuli is present. Other dispositions do not seem to have a specific stimulus to begin with – in that case the C-cases comprise cases of any kind. Talkativity, for instance, is simply the disposition to talk, not the disposition to talk in certain situations. A subject is talkative, therefore, on Manley and Wasserman's account if and only if the subject talks in a suitable proportion of all cases (ibid.: 72).

All cases? Not absolutely all. For apart from the stimulus being present, there are two more important features of C-cases. First, C-cases will have to have the same laws of nature. That seems right. The fact that a thing breaks upon being hit or dissolves upon being placed in liquid in worlds in which the laws are different from ours does not count towards the thing's fragility and solubility, respectively. The actual laws will have to be fixed across the C-cases.

Secondly, C-cases will have to be such that the object retains its intrinsic properties (ibid.: 76). With their focus on intrinsic properties, Manley and Wasserman are in good company – dispositions are usually thought of as a matter of the intrinsic properties of their bearers.[5] Thus, fragility seems to be a matter of a certain molecular structure, and irascibility seems to be a matter of a certain brain structure. And these need to be fixed across the C-cases.

So much for the C-cases. Let's now look at SOPH_D itself in some more detail. Like the simple conditional analysis, SOPH_D analyzes dispositions by means of counterfactuals. However, SOPH_D differs from the simple conditional analysis in one very important way. The simple conditional analysis analyzes dispositions in terms of a *single* counterfactual. If the stimulus occurred, the manifestation would follow. SOPH_D, in contrast, analyzes dispositions in terms of a very large number of counterfactuals with highly specific antecedents, each of which specifies a something close to a complete situation.[6] Thus, to be irascible, say, a person will have to meet a sufficient proportion of counterfactuals of the

---

[5] Recently, a plausible case has been made for the existence of extrinsic dispositions (McKitrick 2003). If that is right, then the C-cases will sometimes comprise extrinsic features as well.
[6] In a critical discussion of the view, Vetter (2011) suggests that there is, in fact, a different interpretation available, but I'll neglect that second interpretation for the time being, since Vetter's suggestion pretty much is to interpret PROP_D in terms of the success view, which is also what I will suggest later on.

form "If C were the case, then the object would freak out", where C will specify a different situation in each counterfactual.

This difference between the simple conditional analysis and SOPH_D is crucial. In contrast to the simple conditional analysis, what matters for SOPH_D is not just what is the case in the stimulus worlds that are overall closest to the actual world. Instead, pretty much *all* possible stimulus cases in which the laws and the object are as in actuality matter. If the counterfactual "If C, then N would M" is true for a suitable proportion of the C-cases, so understood, then the object (or agent) has the disposition. If not, it (or she) lacks it.

How does this solve the problems that beset the simple conditional analysis? Well, the problem of degrees and context sensitivity is solved because, as Manley and Wasserman point out,

> [h]ow big a proportion is 'suitable' will depend not only on the dispositional predicate involved but also on the context of utterance. (...) We can think of the contexts as providing the standards for 'fragility' by establishing a requisite proportion of C-cases in which an object would break, for example. In particular, for something to count as 'fragile' in the mouth of the chemist it must break under a much larger proportion of the sorts of stress conditions relevant to fragility. (ibid.)

This should sound familiar. Manley and Wasserman account for degrees in the very same way as the success view: by means of proportional quantification over cases, where the context determines which proportion of cases is required in a context for the ascription of the disposition to apply *simpliciter* (→ 4.3). Unlike in the case of the success view, the cases go into the antecedents of counterfactuals, of course, but the basic mechanism for degrees and context sensitivity is the same in both cases.

The solution for making cases is likewise familiar. As Manley and Wasserman lay out,

> the right-hand side of our bi-conditional holds even if an object happens to be in 'bad' case where its disposition is (....) masked. All that is required is that the object would break in a suitable proportion of stimulus cases, where these will include any (...) masking cases; and it makes no difference whether these are actual. (ibid.: 76f.)

Here, too, the strategy is basically the same as in the case of the success view (→ 4.6). Since the cases over which the quantification ranges are so diverse that the mask will not be present in most of them, intentions and performances will match up in a suitable proportion, even if a mask is present in the actual world.

When I introduced SOPH_D, I said that the C-cases will be highly specific; they will specify something close to a complete situation. We are now in a position to see why. The presence or absence of any *potential* mask of a disposition has to be settled in the antecedent of each counterfactual. That is why the antecedents of the counterfactuals Manley and Wasserman postulate in their account have to be so specific that they specify, for every potential mask, whether it is present or not. The antecedents of the counterfactuals we obtain that way are highly specific indeed. They specify the situation more or less fully.

Having said all this, let's now return to our original topic: abilities. Obviously, Manley and Wasserman's contribution to the dispositions debate is highly relevant for our concerns in that connection. Let's consider a Manley and Wasserman inspired account of abilities and see how it fares. Here is the view. Call it "the sophisticated conditional analysis of abilities":

> SOPH_A. An agent S has an ability to φ if and only if S would φ in a suitable proportion of C-cases.

What are the C-cases this time? Well, first of all, they will be cases in which an S-trigger is present.[7] That, after all, is the analogue to the stimulus in the case of mere dispositions, and it is crucial to Manley and Wasserman's view of dispositions that the C-cases are stimulus cases. Thus, SOPH_A will have to be read as the view that an agent has an ability if and only if there is a suitable proportion of highly specific C-cases for which the counterfactual "If C were the case, then the agent would φ" comes out true, where each C-case is such that an S-trigger is present.[8]

Note that I am already building an important insight into this interpretation of the C-cases which goes far beyond Manley and Wasserman's template of the view. As we will see later on in this section, proponents of the sophisticated conditional analysis, such as Sosa, think of the C-cases as cases in which the agent is properly *motivated* to φ. But as we know by now, this is unfortunate, because it fails to account for non-agentive abilities.

What more can we say about the C-cases? Again, we have to move beyond Manley and Wasserman's template here. Manley and Wasserman's insight that the C-cases will be cases in which the laws of nature are held fixed surely applies in the case of abilities as well. But that is not the only restriction that will have to be imposed on the C-cases. As we know from our discussion of general and spe-

---

[7] As in the case of the simple conditional analysis, one can in principle tinker with the motivational state in the antecedent. For reasons of unity, I will again focus on intentions.
[8] I am again presupposing that Manley and Wasserman intend their view to be understood in terms of a multitude of counterfactuals and postpone Vetter's (2011) remark that their view may just as well be understood along the lines of the success view.

cific abilities, abilities are had in view of the relevant fact, where it will vary which facts are relevant. Sometimes, we will want to know what an agent can do in view of her total situation; sometimes, in view of her location; sometimes, in view of who is helping; sometimes in view of her intrinsic make-up; and so on.

To accommodate this insight within the framework of the sophisticated conditional analysis, whatever facts go into the "in view of" phrase will have to be fixed across the C-cases. As we know, this means that what is held fixed will vary across contexts and cannot be specified in terms of a single set of rules.

Thus, the C-cases will have to be thought of as cases in which an S-trigger is present, the laws are as in actuality, and the relevant properties of the agent – whatever they may be in a particular context – are held constant. To make this entirely clear, let us refine SOPH_A in the following way:

> SOPH_A'. An agent S has an ability to φ if and only if S would φ in a suitable proportion of the relevant S-trigger cases (the C-cases), where it will vary across contexts which S-trigger cases are relevant.

This, then, is our sophisticated conditional analysis of abilities. It derives from the new dispositionalist assumption that abilities are dispositions, combined with Manley and Wasserman's view about dispositions and supplemented with some of the crucial insights about abilities that we have gained in the course of this book.

Sophisticated conditional analyses have recently been proposed, in different versions, by Sosa (2015) and Vihvelin (2013). Sosa, as I read him, advances something very close to SOPH_A'. Sosa writes:

> What then is required for possession of a competence? Required for archery competence, for example, is a sufficient spread of possible shots (covering enough of the relevant shapes and situations one might be in as an agent) where one would succeed if one tried, an extensive enough range. (Sosa 2015: 98)

As I read this admittedly sketchy passage, what Sosa lays out here is that for an agent to have an ability, the counterfactual "If the agent tried, she would succeed" has to turn out true for a multitude of possible shots which cover various situations in which the agent may be. This is very close to what Manley and Wasserman lay out for dispositions and to what I have suggested as an analogue of their view for abilities.

Note that Sosa is concerned with an *agentive* ability in this passage – the ability to hit a target. Thus, he does not phrase his view in terms of S-triggers. For the example he discusses, that is fine. When it comes to abilities quite generally, the view will have to be stated in terms of S-triggers. Once it is thus reformulated, it just is what I spelled out under the heading SOPH_A' above.

Note, too, that Sosa is quite aware of the fact that the realm of the C-cases will vary across contexts. He postulates that the possible shots will have to cover "enough of the *relevant* shapes and situations one might be in as an agent" (Sosa 2015: 98, my emphasis). And which shapes and situations are relevant? That will vary. To see that, note that Sosa distinguishes between three types of competence:

> (a) the innermost driving competence: that is, the structural seat in one's brain, nervous system, and body, which the driver retains even when asleep or drunk [= *seat*], (b) the fuller inner competence, which requires also that one be in proper shape, i.e., awake, sober, alert, etc., [= *shape*] and (c) the complete competence or ability to drive well and safely, which requires also that one be situated with control of a vehicle, along with appropriate road conditions pertaining to the surface, the lighting, etc. [= *situation*] (ibid: 96f.).

It does not matter for our purposes here whether or not the three types of competence Sosa outlines are exhaustive. What *is* important is that Sosa is very explicit on his view that while a condition like being tied down, say, "does not affect the inner [seat + shape] competence, nor the innermost [seat] competence", it "*does* bear on one's complete [seat + shape + situation] competence" (ibid.: 98).

This is interesting, because, plausibly, the way the view about competence Sosa advances delivers all of those uses of "competence" is by restricting the relevant shapes and situations one might be in as an agent he talks about when he formulates his view on competences as a whole. Sosa is thus well aware of the fact that the C-cases will have to be varied in accordance with the context of an ability ascription.

Vihvelin advances a somewhat different view. Brushing over some of the details, however, it becomes clear that, structurally, she defends something very close to SOPH_A' as well. Vihvelin argues that

> VIHV. S has the narrow ability at time t to do R as the result of trying if and only if, for some intrinsic property B that S has at t, and for some time t' after t, if S had the opportunity at t to do R and S tried to do R while retaining property B until time t', then in a suitable proportion of these cases, S's trying to do R and S's having of B would be an S-complete cause of S's doing R (Vihvelin 2013: 187).

This is quite an unwieldy condition, but for current purposes, we can focus on the structural features of the view. As I read Vihvelin (and I may be mistaken here), Vihvelin's condition is just an enriched version of SOPH_A' with a narrower focus. The focus is narrower because Vihvelin is exclusively interested in what she calls "narrow abilities" – abilities that are held in virtue of an agent's intrinsic properties. The view is an enriched version of SOPH_A', because those abilities are then analyzed in terms of a multitude of rather complex counterfactuals *roughly* of the form

> * If S were in a (relevant?) case in which S had the opportunity and tried and had intrinsic property B, then the trying and B would cause S's doing R.

As in the case of SOPH_A', a sufficient proportion of those counterfactuals will have to come out true. Structurally, then, VIHV is just another version SOPH_A'.

An obvious difference, of course, is that Vihvelin (as I read her) postulates further requirements for a case to qualify as a C-case. Apart from being such that the agent tries to φ in it (or, as I suggest, some S-trigger for φ-ing is present), a case will also have to provide the agent with an opportunity and some intrinsic property of the agent will have to be retained in order for it to qualify as a C-case. But these latter two requirements are just additional layers of complexity and can be neglected for what is to come. Everything I say in what follows applies to VIHV just as much as it applies to the basic version of the view as put forward by Sosa.

So much for the exposition of the sophisticated conditional analysis of abilities. Let's now move on to scrutinizing it. The view is promising – that much is clear. As in the case of dispositions, masks, degrees, and context sensitivity are unproblematic. I don't think we have to go through that in detail. SOPH_A handles these things analogously to the way SOPH_D does, which means that it uses the same basic tools that also play a crucial role in the success view. I trust that the reader has a good understanding of these issues at this stage in the book.

The real problem lies elsewhere. Just like the simple conditional analysis, SOPH_A' runs into the problem of impeded intentions. To yield the right verdict, it should be possible for SOPH_A' to be *not* met in such cases, thus yielding the very prevalent sense in which a coma patient, say, cannot raise her arm. But does SOPH_A' yield this result? Not as far as I can see.

According to SOPH_A', an agent has such an ability if and only if a multitude of counterfactuals of the form "If C obtained, S would φ" are true. "C" specifies one particular C-case in each counterfactual, and each of those C-cases will be an S-trigger case. In the case of an agentive ability like the ability to raise one's arm, it will be an intention case. Thus, what SOPH_A' postulates is that an agent has the ability to raise her arm if and only if a multitude of counterfactuals of the form "If S intended to raise her arm and the case is otherwise relevant (and thus a C-case), S would raise her arm" are true.

But now a problem looms. Either the coma is held fixed across the C-cases or it is not. As I will argue in what follows, SOPH_A' is – wrongly – satisfied in both cases. This may not be obvious at first, because at first sight, holding the coma fixed seems like a very promising move. The problem with the *simple* conditional analysis, after all, was that it always considers the *closest* intention worlds and that those worlds are located *beyond* the worlds in which the agent is in a coma to begin with. The antecedent catapulted us too far out, as it were.

The same does not seem to happen if the coma is held fixed across the C-cases. Since the agent is in a coma in all C-cases, the C-cases will comprise only situations in which the agent is as impeded as in actuality. This promises to yield – rightly – that SOPH_A' will not be satisfied in the coma case. If the antecedent of each counterfactual comprises the fact that the agent is in a coma, then one expects that no consequent will comprise the agent's arm raising. There will therefore not be any true counterfactuals of the form "If C obtained then the agent would raise her arm". Consequently, it seems that SOPH_A' will not be met.

But as good as it sounds, this line of thought is mistaken. In the coma case, the C-cases comprise only coma cases, that much is true. But apart from that, the C-cases have to meet some other criteria, too. Most importantly, they will have to be cases in which the agent intends to ɸ (in the case of agentive abilities). And here we encounter a problem. For as we know, there simply are no such cases to begin with. No coma case is an intention case.

In the framework of the success view, this insight plays a key role in the treatment of cases of impeded intentions. The success view (of agentive abilities) states that the agent has to ɸ in a sufficient proportion of relevant intention cases. In cases of impeded intentions, there are no relevant intention cases, because no coma case is an intention case. And that is good. For if there are no relevant intention cases, then the proportion of arm raising cases among the relevant intention cases is unspecified and *a forteriori* not sufficiently high. The condition postulated in the success view therefore rightly fails to be met in such a case.

The same is not true in the case of SOPH_A', however. For at least on the standard semantics of counterfactuals, a counterfactual comes out vacuously *true* if there are no antecedent worlds (Lewis 1973: 16). Thus, the proportion of C-cases for which it is true that the agent would raise her arm, if that case obtained is actually very high. For all C-cases, the counterfactual will be satisfied. Thus, SOPH_A' yields the verdict that the coma patient *has* the ability to raise her arm. The obvious sense in which she cannot raise it fails to be accounted for.[9]

Vihvelin's more complex version of SOPH:A' is not better off. On Vihvelin's view, recall, the narrow, i.e. intrinsic, ability to raise one's arm is analyzed in terms of a multitude of counterfactuals of the form "If S were in a (relevant?) case in which S has the opportunity to raise her arm and tried to raise her arm and had intrinsic property B, then the trying and B would cause S's raising her arm, where C is a placeholder for various circumstances." Again, the antecedent seems to be impossible. Holding the coma fixed (either in the form of intrinsic property B or in the form of looking only at coma-circumstances), there seem to be no opportunity worlds left.

Or are there? Let's allow for that; an opportunity for φ-ing may simply be a situation in which nothing external conflicts with φ-ing, in which case a coma patient may well have an opportunity to raise her arm. Even then, though, the antecedent will come out impossible. Holding the coma fixed, there will not be any worlds left in which the agent tries to φ. Despite all complexity, the basic problem remains.

Can this consequence be avoided, if we vary the coma? Unfortunately not. First of all, it is entirely unclear why varying the coma should be permitted in the first place. The sophisticated conditional analysis can only deliver an account of general and specific abilities if the context fixes the set of facts that have to obtain throughout the C-cases. Whatever determines the relevant worlds in the framework of the success view goes into the C-cases in the framework of the sophisticated conditional analysis. That's the mechanism one needs, if the distinction between general and specific abilities is to be made sense of. It is hard to see how one could invariably vary the coma without losing the capacity to give a convincing account of general and specific abilities.

Even if the coma (and other impediments of the same kind) were varied, however, this would not solve the problem of impeded intentions. Varying the coma channels us back to the problem of the impeded intention as it arose for the *simple* conditional analysis: when the coma is varied across the C-cases, then there is no reason to assume that the agent would not raise her arm in a suitable proportion of C-cases. The coma would be treated like a mask on that account. While present in the actual world, it is varied across the C-cases. In masking cases, SOPH_A' yields – rightly – that the agent *has* the masked ability. In cases of impeded intentions, this is the wrong outcome. Thus, no matter how we twist and turn the view, impeded intentions remain troublesome for the con-

---

9 Sosa seems to be oblivious to this problem (cf. Sosa 2015: 98).

ditionalist – the increased sophistication of SOPH_A' does not seem to be of much help here.

Yet, that is not to say that SOPH_A' is doomed. For of course, there is one simple move open to the proponent of the view that has also proved advisable for proponents of the simple conditional analysis. The proponent of SOPH_A' can simply supplement her view by adding a further condition that is explicitly designed to avoid the unwelcome consequences in the coma case and other cases of impeded intentions.

To see how, let's return to the first horn of the dilemma once more. The problem here was that, since the C-cases have to be both coma cases and cases in which the agent intends to φ, there will simply be no C-cases to begin with. All counterfactuals of the form "If S intended to φ and the case is otherwise relevant (and hence a coma case), S would φ", will come out vacuously true, because the antecedent specifies an impossible state of affairs. Hence, SOPH_A' fails to account for the very prominent sense in which the coma patient lacks the ability to raise her arm or get up from her bed.

The proponent of SOPH_A' can avoid this problem by adding a supplementary postulate to SOPH_A': the C-cases have to be possible states of affairs in the first place. In cases in which the coma is held fixed across the C-cases, this supplementary postulate is obviously not met. Hence, the so supplemented view yields, rightly, that the coma patient lacks the ability to raise her arm in such cases. (Of course, the coma does not have to be held fixed in *every* context. It may well be varied, in which case the supplemented view yields, again rightly, that the coma patient has the ability to raise her arm. This is all as before. The problematic sense for the original SOPH_A' is just the one in which it is true that the coma patient cannot raise her arm.)

The supplemented view I suggest should replace SOPH_A' is therefore the following:

> SOPH_+. An agent S has an ability to φ if and only if (i) S would φ in a suitable proportion of the relevant S-trigger cases (the C-cases), where it will vary across contexts which S-trigger cases are relevant, and (ii) there are possible S-trigger cases (C-cases) to begin with.

Thus understood, cases of impeded intentions, such as the coma case, are rendered unproblematic. Even though condition (i) of SOPH_+ is met, condition (ii) is not when the coma is fixed across the C-cases.

But then isn't SOPH_+ just as powerful as the success view? After all, it seems to have the very same merits. Is it just a matter of taste which view one adopts? As far as I can see, the answer is a qualified yes. The success view and SOPH_+ both account for the adequacy conditions and they don't run into the problems that proved destructive for the simple conditional analysis or possibilism.

The reason for that, however, is that SOPH_+ *is* the success view in counterfactual guise. First of all, as should have become clear in the previous discussion, the program of the sophisticated conditional analysis only gets off the ground if some key elements of the success view are integrated. And secondly, as will become clear in a moment, once those elements *are* integrated, the two views actually turn out to be almost equivalent.

To see that, recall that the success view states that for an agent to have an ability to φ, the agent has to φ in a sufficient proportion of the relevant possible situations in which the agent intends to φ. SOPH_+ states that for an agent to have an ability to φ, "If C obtained, S would φ" has to be true for a sufficient proportion of C-cases, where those C-cases in turn are relevant possible situations in which S intends to φ.

The only difference between the two views is that the success view postulates that a sufficient proportion of C-cases *are* φ-ing cases, whereas SOPH_+ postulates that a sufficient proportion of C-cases be such that the counterfactual "If C were the case, S would φ" holds for them. At first sight, that seems to make a difference, logically speaking. But insofar as each C-case is a fully specified situation, the two views are in fact equivalent. And as we saw above, the C-cases will in fact have to be (almost) fully specified situations in order for masks to be accounted for.

If that is so, then SOPH_+ and the success view turn out virtually equivalent. To say that a sufficient proportion of the relevant C-cases are such that the counterfactual "If C were the case, S would φ" turns out true is only to say that a sufficient proportion of the relevant C-cases are φ-ing cases. The counterfactual only turns out true for some C-case, after all, if that C-case is a φ-ing case. That's just what happens when you insert whole situations into the antecedent of a counterfactual. You do violence to the counterfactual by turning it into a material conditional.[10]

Speaking of violence, let me now tell you why I think the success view, as I have stated it in this book, is to be favored over the counterfactual rendering of the same idea in terms of SOPH_+. I don't see why one should use counterfactuals in the statement of the view, if the counterfactuals don't actually function as counterfactuals. Nostalgia could be one reason: "The simple conditional analysis

---

[10] As far as the C-cases are not fully, but almost fully specified situations, there is a logical difference between the views, but that difference is minuscule.

looked so promising, so let's see how we can tweak the framework until we eventually get a version that works." Personally, I don't find this motive very convincing, but if you are strongly attracted to counterfactuals (even the somewhat perverted ones figuring in SOPH_+), be my guest and cling to them. If you are rather more of a non-sentimental type, I take it you are better off with the success view as I have suggested it. When it comes right down to it, the counterfactuals in SOPH_+ are a scam. Let's be straightforward. Let's endorse the success view instead.

6.3 Possibilism+

In this section, let us look at a view we get by analyzing abilities along the lines of Vetter's (2015) account of dispositions. This is a fruitful thing to do, because the limitations of the resulting view of abilities will once more shed light on the importance of the success rate as spelled out by the success view.

Vetter's project in her book is to reduce modality in general – including metaphysical modality – to the potentialities of objects and agents. Paradigmatic instances of potentialities, on her view, are dispositions. Even though Vetter does not endorse possible worlds metaphysically, she views possible worlds semantics as useful enough to state an analysis of what it is for an agent to have a disposition along the following lines:

> POSS+_D. An object x has a disposition to φ if and only if x φ's in a sufficient proportion of the relevant possible situations.[11]

Let's call this view "possibilism+ about dispositions". Being soluble, on this view, is for something to dissolve in a sufficient proportion of the relevant worlds; being irascible is for someone to freak out in a sufficient proportion of the relevant worlds.

With this view, Vetter is in obvious opposition to all kinds of views about dispositions which stand in the tradition of the conditional analysis, because Vetter does not believe, and she is very explicit about that (Vetter 2011), that dispositions are individuated, even just in part, by their stimuli. On her view, they are individuated exclusively by their manifestations.

In this respect, she is fully in line with the ideas of Kratzer (1977, 1981) and others, who treat disposition ascriptions in the same way as they treat ability as-

---

[11] Vetter speaks of worlds where I used the term "situations", but what she means by that is just what I mean by "situations".

criptions: as possibility statements. On that view, which we can call "possibilism about dispositions", for a thing to be soluble is for it to be possible for that thing to dissolve. This, in turn, is spelled out in terms of possible worlds: for p to be possible is for there to be a relevant world in which p is the case. So far, nothing new; possibilism about dispositions is analogous to possibilism about abilities as we have discussed it at length in chapter 3.

Of course, Vetter does not endorse possibilism. Possibilism about dispositions faces some of the same problems that possibilism about abilities faces too. Like abilities, dispositions come in degrees and they are context sensitive, for instance (Manley & Wasserman 2008). Possibilism about dispositions is just as unapt to account for those features as possibilism about abilities is.

For this reason, Vetter squares the possibilist framework with the insight that having a disposition is a matter of proportions. In doing so, she incorporates one crucial element that is also part of the success view and of the sophisticated conditional analysis. The degree of an object's disposition to φ corresponds, on Vetter's view, to the proportion of the relevant possible situations, in which the object φ's among the relevant possible situations as a whole. This allows her to account for degrees and context sensitivity.

While degrees and context sensitivity are accounted for by the same means as in the case of the sophisticated conditional analysis and the success view, however, there is one very crucial difference between Vetter's and these other two views. Vetter sticks closer to possibilism in that she clings to her conviction that dispositions are individuated by their manifestation alone.

The sophisticated conditional analysis (→ 6.2) and my own suggestion of how to analyze dispositions (→ 5.4) both put crucial emphasis, not just on the manifestation of dispositions, but on the stimulus as well. The sophisticated conditional analysis looks at a multitude of counterfactuals of the form "If x had been in C, x would have φ-ed", where C-cases are, among other things, cases in which a stimulus is present. The view I suggested, DISPOSITION, looks at the relevant possible situations in which a stimulus is present and evaluates whether or not the proportion of the φ-ing situations among those is high enough.

The same is not true for POSS+_D. POSS+_D restricts the possible situations to the relevant ones – the ones in which the relevant facts obtain – and varies whether or not the stimulus occurs. The relevant facts are facts about the object; usually, its intrinsic properties. Everything else is varied. And while DISPOSITION postulates that the proportion of the φ-ing cases has to be high enough among the relevant *stimulus* cases, POSS+_D postulates that the proportion of the φ-ing cases has to be high enough among the relevant possible situations as a whole. The proportion that matters is therefore a different one.

In what follows, I will not discuss Vetter's account of dispositions. Vetter's book is very rich, and I could not possibly do it justice in the course of this section. I will therefore content myself with a few remarks about the relationship between Vetter's account of dispositions and my own views on abilities and dispositions at the end of this section. Primarily, however, I want to focus on whether Vetter's account of dispositions also provides a viable framework for abilities. As I will argue, it does not.

Applying Vetter's thinking about dispositions to abilities, what we get is the following view:

POSS+_A: An agent S has an ability to φ if and only if S φ's in a sufficient proportion of the relevant possible situations.

Let's call this view "possibilism+ about abilities". Let me point out very clearly right away that Vetter does not endorse this view. In fact, she rejects it, and partly for reasons I am going to discuss in a moment. Yet, POSS+_A is an interesting suggestion, and one worth exploring, because it is so closely related to the success view and will once more shed light on the importance of the success rate – the modal tie between intentions (or other S-triggers) and performances postulated by the success view.

As is easy to see, POSS+_A follows the same scheme as POSS+_D. Having an ability to φ, just like having a disposition to φ, is for the agent to φ in a sufficient proportion of the relevant possible situations. No mention of intentions (or any other S-trigger). Analogously to the case of POSS+_D, it will vary, across the relevant possible situations, whether or not an intention is formed. What one checks for is the proportion of the φ-ing cases to the relevant possible situations as a whole.

The relevant possible situations, of course, are the situations in which the relevant facts are held fixed, and as we know from our discussion of general and specific abilities, it will have to vary which facts are relevant in a given context. Sometimes, we hold only stable, mostly intrinsic facts about the agent fixed; sometimes, we hold lots of facts about the external situation fixed as well. We do not, however – and this is very important when it comes to the difference between POSS+_A and the success view – restrict the relevant possible situations to the ones in which an intention is formed.

What we have here, then, is a view that resembles the success view very closely. Subtract the restriction to intention situations from the success view and what you end up with is POSS+_A. That makes POSS+_A a very interesting view to look at in some detail. If POSS+_A comes out right, then the success view would turn out unnecessarily complex. Why bother with intentions and the success rate, if focusing on the ratio of performances among the relevant possible situations as a whole does the job just as well?

As we will see in what follows, however, POSS+_A does not come out right. There is good reason for focusing on the success rate and not, as POSS+_A suggests, on the ratio of performances to the totality of the relevant possible situations. For simplicity, let us focus on agentive abilities in what follows.

First of all, I take it to be highly unintuitive to look at the proportion of ϕ-ing cases in relation to all relevant possible situations, given that the agent will not intend, not have reasons, and not desire in any way to ϕ in the vast majority of those relevant possible situations. Think of my general ability to hit a particular bull's eye; one in a bar in Sweden, say. I have that ability. I have it in view of a variety of my features. Let's focus on the ability in view of one set of features: my brain structure and muscular constitution. Now, according to POSS+_A, I have the ability to hit that bull's eye in Sweden, because I hit it in a sufficient proportion of situations in which my brain structure and my muscular constitution is as in actuality.

But just think of how incredibly small that proportion is going to be! In the vast – let me say it again: vast! – majority of all situations in which those features are as in actuality, I don't have the slightest intention or reason to go about hitting the bull's eye in a bar in Sweden. In fact, I will not be in Sweden in most situations and will not have the slightest reason or intention to hit the bull's eye in that bar. POSS+_A, however, factors all of those situations in when it comes to determining whether or not I have the ability to hit that bull's eye. It looks at the proportion of hitting situations in relation to all situations in which my brain structure and muscular constitution is as in actuality.

Now think about differences in the degree of an ability. According to POSS+_A, someone with a high degree of the ability under consideration and someone with a low degree of that ability differ with respect to the proportion of hitting cases among the situations in which their brains and muscles are as in actuality. But given how minuscule the proportions are in *each* case, the difference will be almost unspeakably tiny.

The same is true for POSS+_D, by the way. Take solubility. Sugar has that disposition in view of its molecular structure. According to POSS+_D, this is because there is a sufficient proportion of dissolution cases among the relevant possible situations – the situations in which sugar has its actual molecular structure. Note, however, that in the vast majority of those situations, the sugar is not placed in any kind of liquid. Thus, the proportion of dissolving cases among the relevant possible situations will be minuscule, and so will differences between highly soluble, moderately soluble, and almost insoluble substances.

This is not a knock-down argument against POSS+_D or POSS+_A. But it raises doubts as to whether the proportion the views postulate are really what matters. In what follows, I will focus exclusively on POSS+_A and argue that the proportion it postulates is certainly not the right one. Consequently, POSS+_A fails. As we'll see, the reasons for that failure speak very much in favor of the success view.

That the proportion POSS+_A postulates is the wrong one can be seen by looking at a case Vetter herself puts forward against POSS+_A (ms.). Take an obsessive-compulsive agent, whose intrinsic psychological constitution is such that the agent forms the intention to wash her hands every five minutes. Compare that agent to a psychologically sane person, who only washes her hands in standard situations. We can assume that both agents are equally good at washing their hands – they have the hand washing ability to the same degree. Their psychological constitution does not make a difference in this regard.

The problem is that POSS+_A predicts otherwise. That is because, holding each agent's psychological constitution fixed, the proportion of hand washing situations is going to be much higher in the case of the obsessive-compulsive agent than in the case of the sane agent. And hence, the obsessive-compulsive agent should come out as having the ability to a higher degree. (The example may not be ideal; degrees of the ability to hand wash are usually almost exclusively evaluated on the level of achievement. To see that this does not affect the argument, though, I invite you to run the same argument with an obsessive-compulsive dart player instead.)

What does the argument show? First, of all, I submit, it shows that there is something wrong with POSS+_A. Perhaps you are not willing to draw that conclusion right away. Perhaps you are inclined to say "Wait a minute. Why can't the proponent of POSS+_A simply resist your move of holding the psychological constitution of the agent fixed? The obsessive-compulsive agent does not have the ability to wash her hands to a higher degree than any average person, because each agent's ability to wash their hands has nothing to do with their respective psychological constitutions. Why should the proponent of POSS+_A allow for you to fix an arbitrary set of features to evaluate whether or not the agent has the ability to wash her hands?"

In response, let me first of all clarify one thing. I fully agree that it is not legitimate to fix an arbitrary set of features to evaluate whether an agent has some ability. It is not, for example, legitimate to fix my broken arm in evaluating whether or not I have the ability to do a handstand in view of my muscular constitution. Likewise, it is not legitimate to fix the fact that I am asleep in evaluating whether or not I have the ability to sing in view of my larynx system. What *is* legitimate, however, is to fix my broken arm in evaluating my ability to do a handstand in view of my broken arm, and to fix the fact that I am asleep in evaluating my ability to sing in view of my being asleep.

Casting a look back at the argument against POSS+_A, you will note that the ability under consideration there is the ability to wash one's hands *in view of one's psychological constitution*. To evaluate whether or not and to what degree an agent has that ability, we are not only allowed to, but we actually *have* to hold the agent's psychological constitution fixed. And doing so, we get wrong results. Thus, there is something wrong with POSS+_A.

But perhaps it is illegitimate to look at the ability to wash one's hands in view of one's psychological constitution from the outset? What kind of ability is that supposed to be after all? Shouldn't the proponent of POSS+_A resist focusing on weird abilities like that right from the outset? Differently put, shouldn't she resist the idea that any set of facts go into the "in view of" phrase?

In response to this way of framing the objection, I would like to stress two points. First of all, I really do think, and I wish to commit myself to this with respect to the success view as well, that any set of facts *can* go into the "in view of" phrase of any ascription of the form "S can, in view of F, ϕ". Our view should yield the right results, no matter whether we look at the ability to sing in view of one's larynx system or the ability to sing in view of the Chinese economy crisis or in view of the amount of hair on my head or in view of the fact that your favorite color is red and I have never been in Korea. There is nothing wrong with those kinds of abilities. They are weird, all right. But there are no good grounds for rejecting them. Facts that don't have anything to do with an ability should be safely insertable into the "in view of" phrase and not make any difference to the agent's having or lacking them. This is not the case when it comes to POSS+_A.

The second, and perhaps even more important point is that the ability to wash one's hands in view of one's psychological constitution is not a "weird" ability to begin with. It is in view of her psychological constitution, after all, that someone phobic of water does not have the ability to wash her hands. She does have that ability in view of many other features: in view of her strength and motor control she has it, for instance. But there is a very clear sense in which she lacks the ability to wash her hands. To account for that sense, we need to be looking at her ability to wash her hands in view of her psychological constitution. The ability to wash one's hands in view of one's psychological constitution is a perfectly legitimate ability. I thus take it that the argument above really does show that something is wrong with POSS+_A.

But what exactly is wrong with it? I submit the trouble stems from the fact that the proportion of hand washing situations to the totality of relevant possible situations is not the proportion that matters. What really matters is the proportion of hand washing situations to the relevant possible situations in which the agent intends to wash her hands. And that is just what the success view states. Thus, the case speaks very much in favor of imposing an additional restriction to intention situations upon the relevant ones as a whole.

The success view can explain very smoothly why the obsessive-compulsive agent is not better at washing her hands than any average agent: her success rate is not higher. The ratio of situations in which the obsessive-compulsive agent washes her hand among the relevant possible situations in which she intends to do so is just about the same than the same ratio in the case of the sane person. This bolsters my claim that it is the modal success rate which matters when it comes to (degrees of) abilities. Without the further restriction to intention situations, counterexamples arise.

At least in the case of abilities, then, there is strong reason to think that an individuation solely in terms of the manifestation – or, rather, the exercise – is on the wrong track. Abilities, it seems, have to be analyzed by a ratio which ties intentions (or S-triggers quite generally) to exercises. The success view postulates just that.

In doing so, the success view does not run into the unintuitive consequence of minuscule proportions either, by the way. The proportions the success view postulates are usually quite high. Typically, an agent who is good at something will more often than not succeed in doing that thing upon intending to do it. Within the framework of the success view, minuscule proportions are the exception, not the rule, if someone has an ability.

6.4 An option-based account

Maier (2013) is concerned with a certain class of properties, which he refers to as *the agentive modalities*. The agentive modalities, as Maier understands them, include options, abilities, skills, and potentialities. They are modal properties of some agent, which involve some relation to the agent, and in virtue of which certain "can" statements are made true. Agentive modalities are the truth makers of statements of the form "S can A" where "S" denotes the agent and "A" refers to some act-type (Maier 2013: 2).

As Maier lays out, there are two different kinds of questions one may ask about agentive modalities. On the one hand, there are what he calls external questions – questions about the relationship between agentive modalities and other modal properties, such as the question whether or not agentive modalities are reducible to other modal properties or whether the metaphysics of modality will "require a species of modality that is distinctively agentive" (ibid.). On the other hand, there are what he calls internal questions, which concern the relationship between agentive modalities themselves. "Are they miscellany, or a unified collection? If the latter, what is it that unifies them?" (ibid.).

With respect to external questions, Maier remains officially neutral (ibid.:10), although he formulates strong reserves concerning a reductive enterprise (ibid.:7). What he is explicitly concerned with are the internal questions. Here, Maier offers an account, according to which all agentive modalities can be elucidated in terms of the most basic among them. Maier calls them *options*. He writes:

> There are at least two ways in which a given agent may be able to perform a certain action. It may be now in her power to perform that action: there is, as it were, nothing between her and the deed. This corresponds to what I have been calling options. Alternately, she may be able to perform the action, though it is not now in her power to do so. For example, Rachel may be able to speak Mandarin even though, under the effects of novocaine, it is not now in her power to do so. Let us call this latter way of being able to perform an action a general ability. (Maier 2013: 11)

The set-up of the example should ring a bell: options seem to be specific abilities.[12] But they seem to be specific abilities of a special kind. As Maier has it, options are "the species of agentive modality that figures in the framing of choice situations" (ibid.:8). This makes options particularly well-suited as a fundament for an account of agentive abilities, in Maier's view, because they have two important features: they are "[d]eterminate in the sense that it is perfectly clear, in some given case, whether or not this agentive modality obtains" (ibid.:7). And they are "[i]ndispensable in the sense that, in our theorizing about agency, this agentive ability is one that we cannot do without" (ibid.:7).

Why think that options have those two features? According to Maier, they have them in virtue of their role in decision theory. Options are indispensable, because questions of rationality or morality cannot be formulated without them. In order to be able to ask what some agent ought to do, we need to compare the (moral or rational) value of her options (ibid.:8f.). And they are determinate, because otherwise the recommendations of decision theory could not be determinate. Since recommendations like "Take the option that maximizes overall utility" *are* perfectly determinate, however, so has to be the notion of an option (ibid.: 9).

From here, it is only a small step to Maier's full account of general abilities. For general abilities can now be defined in terms of options:

---

[12] Maier has confirmed this in personal conversation.

> MAIER. S has the general ability to A if and only if S has, in a suitable proportion of similar circumstances, the option of A-ing. (ibid.: 19)

This yields the following picture. On the one hand, there are options which play an indispensable role in moral theory and decision theory and are well-understood by their theoretical role in the framing of choice situations. Options are presumably basic in the sense that their modality cannot be reduced any further. On the other hand, there are general abilities which are defined in terms of options. To have a general ability is to have an option in a suitable proportion of similar circumstances.[13]

So much for the exposition of Maier's view. Let's now see how the view relates to the success view and what to make of it. Some similarities between Maier's account and the success view should be obvious. First, Maier explicitly endorses Kratzer's insight that abilities are always had in view of certain facts. That is why he restricts the possible circumstances to the similar ones. We obtain the similar circumstances by holding certain facts fixed. Which facts we hold fixed can vary. In this respect, Maier's view resembles the success view rather closely.

Yet, I take the success view to be one step ahead. The success view speaks of *relevant possible situations* where Maier speaks of *similar* circumstances. The reason is that we may sometimes be interested in what the agent can do given certain *hypothetical* circumstances. As I have laid out earlier (→ 4.5 f) the relevant possible situations will in that case contain some that are *dissimilar* to the actual world in crucial respects. Maier's account fails to provide the resources for cases like that. The success view is more flexible in that respect.

The second obvious similarity between Maier's view and the success view is that both analyze abilities in terms of the notion of a suitable proportion of circumstances. The proportions that matter are different, of course. Maier counts the options among the similar circumstances, while the success view counts performances among the relevant intention cases, but the general idea of applying proportional quantification to cases is obviously the same. Moreover, Maier does not work with the idea of a weighted proportion, but obviously he can avail himself of that tool as well. Basically, the success view and Maier's view are on a par here.

The merits of the proportional quantification are the same in both cases, too. Proportions deliver an account of degrees, and the ability *simpliciter* is had if and only if the degree of the ability is high enough. Thus, an agent's degree of an ability is higher, on Maier's view, the higher the proportion of option cases

---

[13] The class of the agentive modalities is wider. I will focus exclusively on abilities, though.

among the similar circumstances is. Apart from the fact that Maier is interested in a different proportion, this looks a lot like the account of degrees I developed in 4.3.

So much for the similarities between Maier's and the success view. Let's now turn to the crucial difference: the central role of options within Maier's framework. According to Maier, what matters for an agent's ability to ɸ is the proportion of the agent's options to ɸ among the similar circumstances. Options, in turn, are defined functionally. They are "the species of agentive modality that figures in the framing of choice situations" (ibid.:8). Here, Maier's view encounters some severe problems.

First, and most importantly, I don't believe the notion of an option is actually as well-understood as Maier seems to think. In telling us that options are "the species of agentive modality that figures in the framing of choice situations" (ibid.:8), Maier makes it seem as though this would settle the issue once and for all. In reality, though, choice theory knows of various ways to define options, some of which are obviously not what Maier has in mind.

Choice theory is very often understood as a psychological model, or as Wolfgang Schwarz puts it, "as part of an idealized, high-level, computational design" (ms.: 1). Such an understanding of choice theory does not seem to yield the fully determinate notion of an option Maier is after to begin with. That is because

> [s]uch a model (...) could be a descriptive model, trying to capture, to a first approximation, the choices made by actual humans. Or it could be a normative model, specifying the choices people should ideally make. Or it could be a constitutive model, implicitly defining what it is to be an agent with such-and-such beliefs and desires. On each interpretation, we can't assume that the available acts are simply given as part of a well-defined decision problem. Somehow the agent's cognitive system must itself figure out the available options. (ibid.:1)

The psychological model of choice theory does not seem to be what Maier needs, then. Instead, options are supposed to be something objective. The agentive paths that are objectively open to the agent at a given time, or something along those lines.

I am not sure that this notion is as clear and well-understood as Maier seems to. Schwarz, at any rate, puts a lot of effort into spelling out what it is to have an option in the objective sense. And what he says will not please Maier, because, first, it does not treat the modality of options as modally fundamental, and, as we will see shortly, it does not sit well with Maier's ideas about reducing abilities to options. Schwarz writes that we could define objective options (for an agent at a time) as things the agent would do if she intended (at the time) to do them. I prefer a slightly weaker alternative that retains the existential form of our first stab: A is an objective option if some variation of the agent's intentions (at the time) would bring about A. (Schwarz, ms.)

In a nutshell, the view seems to be this:

> $CA_{OPTIONS}$. S has an option if and only if there is some variation of the agent's intentions V, such that, were V to occur, it would bring about ϕ.

Of course, the careful reader will immediately balk at this: what about coma patients? What about phobics? Doesn't $CA_{OPTIONS}$ yield wrong results in those cases? If the coma patient intended to raise her arm, she would. If the phobic intended to touch the spider, she would. Hence, $CA_{OPTIONS}$ seems to yield that the coma patient has the option of raising her arm the phobic has the option of touching the spider, respectively. But that seems wrong. Those agents don't seem to have those options.

Schwarz is well aware that demurring statements like this will be put forward by those engaged with abilities. But he does not think they matter to his project. Instead, he points out that his view of options will not be "adequate as a general analysis of the English auxiliary 'can'" (ibid.). But that is okay: decision theory is a completely different business, on his view. Coma patients, for instance, simply fall outside the scope of choice theory altogether:

> decision-theoretic models describe the workings of an idealized cognitive system. They don't apply to rocks or corpses or people in a coma. Decision theory is not meant to predict or evaluate the behaviour of these systems, so we may well set aside the question how to delineate their options."(ibid.)

And phobics? They do actually *have* the option to do what they are phobic of on Schwarz's account. There is what he calls a "disconnection between decision-theoretic options and everyday normative attitudes towards people's actions" (ibid.). Thus, when we are considering what agents can do, in terms of an everyday sense of "can", "we often hold fixed various facts that should not be held fixed when we catalogue the space of decision-theoretic options" (ibid.), according to Schwarz. For instance,

> we exclude options that would go against deeply ingrained fears or hopes or convictions (...). Perhaps we can even hold fixed quite commonplace intentions: since you intend to change lanes, you ought to accelerate (...). There is also a temptation to hold fixed the past and the laws of nature, which is why some people intuit that determinism renders praise and blame pointless: you couldn't really have done anything else. In these respects,

> the range of options that figure in everyday normative evaluations is often more narrow than decision-theoretic options. (ibid.)

This is bad news for Maier. Apparently, it is far from clear that options, as they figure in choice theory, are the kinds of animals to which abilities can be successfully reduced. If a phobia does not deprive an agent of the option to touch a spider, because we don't hold the agent's deeply engrained fears fixed, then it is hard to see how it can possibly deprive her of the general ability to touch a spider, on Maier's view. Holding the phobia fixed, the agent will still have the option to touch spiders pretty much all the time. Thus, Maier's view yields that she also has the general ability. That is obviously mistaken. As far as I can see, the disconnection between choice theoretic options and abilities more generally, which Schwarz points to, is a genuine problem for Maier's account.

There is another problem with using the choice theoretic notion of an option as a fundament of an account of abilities more generally. The notion of an option, as it figures in choice theory, is such that it automatically renders having multiple options compatible with determinism. Witness Schwarz again:

> the kinds of choices studied in decision theory (...) do not require a strong, libertarian kind of freedom. Decision theoretic algorithms are widely used in artificial intelligence, where it is taken for granted that the (artificial) agent's actions can be predicted from its internal architecture and the inputs it receives. If you build a robot, you don't need to specify that it should drop towards the Earth when it falls out of a helicopter. It will do that no matter what internal states and decision rules you build in. Not so when the robot reaches a junction. Here the outcome is under control of the robot's internal state: vary the state, and the robot will choose different paths. That is all the freedom we need. (ibid.)

What Schwarz points out in this passage is that options can be had, even if the actual future behavior of the system is determined. That is a problem for Maier because there seems to be a clear sense in which agents cannot do otherwise, if determinism is true. If all there is to abilities is options in varyingly large sets of circumstances, then no such sense can be spelled out. Holding determinism fixed, there will still be loads of options in the choice theoretic sense Maier uses in his account. And thus, there is no sense in which a predetermined path deprives agents of their abilities to act otherwise.

This is particularly problematic because the quest for a viable account of abilities has always been motivated, in part, by the more overarching quest for a more thorough understanding of the free will debate, with the compatibility problem at its center. There is obviously a sense in which agents cannot act otherwise in a deterministic world, and a sense in which they can. Which sense is relevant to the free will problem? This question is one that Maier will not be able to tackle with his account of abilities, which already presupposes compatibilism.

Finally, let me note a third problem with an understanding of general abilities in terms of options. It is hard to see how to integrate non-agentive abilities into the framework. I surely have the ability to smell, but I am not sure there are any situations in which I have smelling as an option in the sense in which options figure in choice theory. Options, I should think, are essentially options to perform *actions*. Non-agentive abilities are by definition abilities to exhibit behavior that does not qualify as actions, though. There seems to be a tension here, to say the least.

Of course, Maier will say that his view is intended as an account of agentive abilities only. And that response is perfectly respectable, of course. But, and here comes the qualification, the response is respectable only as long as we can think of a principled way of extending the framework also to cover non-agentive abilities.

In the case of the success view, the account of agentive abilities turned out to be a special instance of a broader scheme, which covers both agentive and non-agentive abilities. In the case of Maier's view, in contrast, it is not too easy to see what such an extension might possibly look like. If (i) the modality of options is irreducible and basic, (ii) Maier's account of agentive abilities is built upon those basic modal properties and (iii) options are the wrong kinds of animals to feature in an account of non-agentive abilities, then an account of non-agentive abilities will inevitably feature different modal properties as their fundament. Agentive and non-agentive abilities will then differ in the kind of modality that underlies them. That is a genuine problem, since the challenge is precisely, I take it, to come up with a view that accounts for the fact that the modality of agentive and non-agentive abilities seems to be of the *same* kind.

We can note, then, that Maier's account faces some severe difficulties, even though it obviously shares some important features with the success view. The problematic part, I have argued, is the attempt to reduce abilities in general to the choice theoretic notion of an option. Doing so plausibly yields wrong verdicts about phobics and other psychologically impaired agents, it fails to yield the obvious sense of ability, according to which an agent is unable to act otherwise, if determinism is true, and it yields problems when it comes to the integration of non-agentive abilities. I take it that these are three strong reasons to try and get along without the choice theoretical notion of an option.

6.5 Upshot

In this chapter, we situated the success view within the contemporary literature on abilities and among some potential views that result from mapping some of the most promising recent contributions in the literature on dispositions onto the ability realm. As we saw, all of the recent accounts operate with the notion of proportions of worlds or situations. And most of them (Maier's option-based view excluded) spell out the same idea that also underlies the success view: that having an ability is a matter of a sufficient performance rate across a portion of the modal realm. In their broad outlines, then, the success view and the other views that were discussed in this chapter do not differ too much. As always, though, the devil is in the details. Looking more closely at each of the views, their more subtle relations to the success view become apparent.

Greco's (2007) view, I argued in section 6.1, is fully compatible with the success view, in that it states, roughly, that abilities are a matter of success across a sufficient proportion of the relevant possible situations, which is the core idea of the success view as well. What Greco does not say is how this is to be spelled out; here, the success view adds an important layer. According to the success view, success across a sufficient proportion of the relevant possible situations is for S-triggers of performances and the corresponding performances to match up across a sufficient proportion of the relevant possible situations. Greco, I said, can thus be seen as a 1.0 version of the success view (but also of other views, which can be seen as attempts to spell out what success across a sufficient proportion of the relevant possible situations amounts to).

In section 6.2, we looked in detail at what I called the sophisticated conditional analysis, versions of which have been proposed by Manley and Wasserman (2008) for dispositions, and by Sosa (2015) and Vihvelin (2011) for abilities. According to the sophisticated conditional analysis, having an ability is a matter of a sufficient proportion of counterfactuals of the form "If C were the case and S intended to ϕ, S would ϕ" turns out true, where C picks out a highly specific set of circumstances.

The sophisticated conditional analysis is very similar to the success view in many ways. So similar, in fact, that once it is ameliorated in such a way that cases of impeded intentions can be accounted for, it really just *is* the success view in counterfactual guise. However, and this is really important if you have strong inclinations towards a counterfactual account of abilities, there is good reason to favor the non-counterfactual formulation of the success view. The counterfactuals figuring in the sophisticated conditional analysis may look like counterfactuals all right. Really, though, their antecedents are so specific that they pick out particular situations. Thus, the counterfactual appearance of the sophisticated conditional analysis therefore turns out to be a scam on a closer look and certainly not the most straightforward way of spelling out the thought that is supposed to be captured in the view.

While sections 6.1 and 6.2 dealt with views that can be seen as allies of the success view, I used sections 6.3 and 6.4 to discuss problems with two other views that seem rather appealing at first.

In section 6.3, I looked at a view which is obtained by analyzing abilities along the lines of Vetter's (2015) account of dispositions. The resulting view, which I called possibilism+, resembles the success view closely in that the only difference between the two views is that possibilism+ drops the restriction to intention situations that figures so prominently in the success view. According to possibilism+, having an ability to φ is a matter of φ-ing in a sufficient proportion of the relevant possible situations, full-stop. No further restriction to intention situations is postulated.

The view looks appealing at first, because it is obviously somewhat simpler than the success view. As we saw, however, the view runs into trouble with cases in which agents exhibit an increased tendency to φ in virtue of an increased tendency to intend to φ. The additional layer of complexity provided by the success view is therefore highly justified. Abilities really seem to be a matter of the modal tie between intentions and performances and not just of the sheer proportion of performance cases as such.

In section 6.4, finally, we looked at Maier's (2013) option-based account of abilities. On Maier's view, having an ability is a matter, not of φ-ing, but of having the option to φ across a sufficient proportion of the relevant possible circumstances. Options are irreducible, on his view, but well-understood in virtue of their role in choice theory. I have laid out multiple problems with this view.

First, the notion of an option is not obviously as well-defined as Maier seems to think. Secondly, it is not clear that an account of abilities in terms of options yields the right verdicts in cases of impeded intentions. Thirdly, Maier's view seems incompatible with there being a good sense of having the ability to act otherwise, according to which that ability is incompatible with determinism. Finally, the view does not sit well with non-agentive abilities. I take these points to show that Maier's option-based view of abilities is inferior to the success view in some important ways.

7 The success view applied – Two rough sketches

As laid out in the introduction, I view this book, in part, as groundwork for the more systematic and informed exploration of various philosophical questions. To give you a flavor of how this may go, I want to close by sketching, very roughly, two applications of the success view to well-taken philosophical problems. One is the grandfather paradox, as it arises in connection with time travel. The other is the free will problem.

Why am I choosing those two problems and not any other? After all, several other philosophical problems seem to be just as well-suited to be re-considered in the light of the insights that the success view provides about abilities. We know that virtue epistemologists, such as Greco (2007) and Sosa (2015) analyze knowledge in terms of intellectual abilities. Certainly, it would be quite interesting to look at those abilities more closely against the background of the success view. The same goes for other theories which feature the notion of abilities at center stage, such as Millikan's theory of substance concepts (2000), Mayr's theory of action (2011), and various views of conceivability (Yablo 1993; Menzies 1998), to name but a few.

Why focus on time travel and free will? First, of course, because I have to make *some* choice. Secondly, I take it that time travel and freedom are particularly interesting to look at. The time travel topic is interesting because the standard solution to the grandfather paradox appeals to possibilism – a view I have discussed in great detail (→ 3). It is therefore easily accessible against the background of this book and gives me the opportunity to take up a few threads that have been running through the previous chapters.

The reason for talking about free will in the outlook is that the debate about free will seems to me to be among the debates which have been the most strongly affected by a prevailing lack of understanding of abilities. That is because the debate has been dominated, in large part, by the idea that freedom requires agents to have the ability to act otherwise (Frankfurt 1969), but no one seems to know how to understand this requirement. Even worse, some of the most established players in the debate seem to be completely oblivious to the fact that there could be open questions about the understanding of that requirement.

Van Inwagen writes the following in response to the idea that compatibilists and incompatibilists about freedom and determinism may perhaps be talking past each other when fighting about whether or not the ability to act otherwise is or is nor compatible with determinism:

> I want to make what seems to me to be an important point [...]: compatibilists and incompatibilists mean the same thing by 'able.' And what do both compatibilists and incompatibilists mean by 'able'? Just this: what it means in English, what the word means. (van Inwagen 2008 : 333)

To the reader of this book, it should be obvious that this stance towards the meaning of "able" is ignorant to an almost amusing degree. Ability statements, and this much is clear, are highly context sensitive. Hence, worries about whether compatibilists and incompatibilists may not in fact talk past each other obviously cannot be settled by pointing out that they both mean "what the word means" when they talk about abilities.

In my outlook on the applicability of the success view to the free will issue, I will try and push this very point. More specifically, I will use the insights of this book to come up with a contextualist view of freedom statements, which allows for a conciliatory perspective on the ongoing debate between compatibilists and incompatibilists about freedom and determinism. That is not to say that this is the only reasonable application of the success view to the free will problem, but it is a radical one and one worth exploring.

7.1 A paradox about time travel revisited

Let's start with the grandfather paradox. The paradox is this: if time travel were possible, then it would be possible for a situation to occur in which some agent can and cannot kill his own grandfather. Since that it impossible, time travel is impossible.

Here is Lewis's (1976: 149 f.) exposition of the paradox. Suppose Tim travels back to the year 1921 to kill his grandfather. Back in the past, he buys a rifle, trains to shoot, tracks his grandfather down, and aims. Can Tim kill his grandfather? The answer seems to be: yes and no. On the one hand, he can; he clearly has what it takes, and the circumstances are just perfect. On the other hand, he cannot; if he could, he could do what is logically impossible. Tim obviously did not kill Grandfather in 1921, after all, because Grandfather lived on after that and, among other things, fathered Tim's father. If Tim killed him in 1921, it would therefore be true that he killed him in 1921 and did not kill him in 1921. Hence, it is logically impossible for Tim to kill Grandfather in 1921.

Is time travel incoherent, then? Not according to Lewis. If Lewis is right, then the grandfather paradox is merely apparent, because it equivocates on "can". How so? Here, we need to take a step back and take a look at Lewis's views about ability statements.

As we know, Lewis is a possibilist. That is to say that, on Lewis's view, ability statements – and "can" statements quite generally – express restricted possibility claims: "S can ɸ" is true, according to Lewis, if and only if S's ɸ-ing is compossible with the relevant facts. In possible worlds terminology that is to say that

> POSS. "S can ɸ" is true if and only if there is a relevant possible world in which S ɸ's, where the relevant worlds are the worlds in which the relevant facts obtain.

We also know that possibilism is a contextualist view about ability statements. Which facts are relevant will vary across ascriber contexts. Recall Lewis's famous quote on that issue:

> To say that something can happen means that its happening is compossible with certain facts. Which facts? That is determined (...) by context. An ape can't speak a human language – say, Finnish – but I can. Facts about the anatomy and operation of the ape's larynx and nervous system are not compossible with his speaking Finnish. The corresponding facts about my larynx and nervous system are compossible with my speaking Finnish. But don't take me along to Helsinki as your interpreter: I can't speak Finnish. My speaking Finnish is compossible with the facts considered so far, but not with further facts about my lack of training. What I can do, relative to one set of facts, I cannot do, relative to another, more inclusive, set. (ibid.:150)

Possibilism plays a crucial role in Lewis's solution to the grandfather paradox. That is because, on Lewis's view, the formulation of the paradox equivocates on "can" in that the set of facts relative to which Tim can kill his grandfather is different from the set of facts relative to which he cannot:

> Tim's killing Grandfather that day in 1921 is compossible with a fairly rich set of facts: the facts about his rifle, his skill and training, the unobstructed line of fire, (...) and so on. (...) Relative to these facts, Tim can kill Grandfather. But his killing Grandfather is not compossible with another, more inclusive set of facts. There is the simple fact that Grandfather was not killed. (...) Relative to these facts, Tim cannot kill Grandfather. He can and he can't, but under different delineations of the relevant facts. You can reasonably choose the narrower delineation, and say that he can; or the wider delineation, and say that he can't. But choose. What you mustn't do is waver, say in the same breath that he both can and can't, and then claim that this contradiction proves that time travel is impossible. (ibid.: 151)

Here is Lewis's solution to the grandfather paradox in a nutshell:

> SOLUTION. The formulation of the paradox equivocates on "can" and the paradox is therefore merely apparent. That is because Tim can kill Grandfather in one sense, but not in another, in that he can kill him, relative to one set of facts, but cannot kill him, relative to another.

SOLUTION seems to be widely accepted throughout the literature on time travel.[1] Yet, of course, it is ill-justified as it stands. The reason for that should be obvious. As I have shown at length in chapter 3, possibilism is false; ability statements are not plausibly understood as restricted possibility claims. Thus, SOLUTION relies on a false presupposition. And as a consequence, it stands on very shaky ground.

What I want to show in what follows, however, is that the same solution flows out of an ameliorated view of abilities – specifically: the success view of ability. Lewis's solution to the grandfather paradox is therefore on the right track, despite the fact that the view of abilities which underlies it is mistaken.

To see this, note that possibilism contains two elements: a restriction of the possible worlds to the ones that are relevant in a context, and the postulate that there be at least one among those worlds in which the agent φ's. As we know, there is something wrong with the latter of those elements. There being at least one relevant φ-ing world is often not enough for an agent to have an ability, and it certainly does not provide the scale of reliability one needs to account for a certain kind of context sensitivity that attaches to ability statements. That is why we need to think of abilities in terms of a modal success rate. There has to be, not one world, but a sufficient proportion of φ-ing cases among the set of cases that matter.

Where possibilism gets things exactly right, though, is with respect to the first element: the restriction of the possible worlds, or situations, to the relevant ones. As we know, abilities are always had in relation to certain facts, and which facts matter will be a matter of context. This element of possibilism should be retained in a comprehensive view of abilities and is therefore part of the success view as well.

Looking back at Lewis's solution to the grandfather paradox, it is easy to see that it is the restriction to worlds that does the work in Lewis's argument. Tim can kill grandfather in relation to one set of facts, but not another. He can kill him in view of his amount of shooting training, the fact that he holds a gun to grandfather's face, and so on. But he cannot in view of the fact that grandfather lived on, in view of the fact that grandfather later fathered Tim's father, and so on.

Of course, this is then spelled out in terms of the wrong framework. According to Lewis, Tim can kill grandfather in view of him holding a gun to his face, because his killing grandfather is *compossible* with Tim holding a gun to his face. And Tim cannot kill grandfather in view of the fact that grandfather

---

[1] But see Vranas (2009) for criticism of this response.

lived on, because his killing grandfather is *not compossible* with grandfather living on. But compossibility is not essential to the argument. The kind of quantification that is applied to the relevant worlds is only secondary here. What really does the work is the restriction to the relevant worlds.

And because this element is retained in the success view, the success view allows for the same take on the grandfather paradox. Tim can kill grandfather relative to the fact that he holds a gun to his face, *because he kills him in a sufficient proportion of cases in which he intends to kill him and holds a gun to his face*. And he cannot kill him relative to the fact that grandfather lived on, *because he does not kill him in a sufficient proportion of cases in which he intends to kill him and grandfather lives on*.

What this shows is that Lewis's solution to the grandfather paradox stands, despite the wrong account of abilities it was originally built upon. And the way this was shown is by applying some of the main findings of this book to the line of reasoning that has led to Lewis's solution and refine it accordingly.

7.2 Alternate possibilities contextualism about freedom

Let me now sketch a second way in which the success view of ability can be put to work. As I will argue in what follows, it offers a fresh, albeit idiosyncratic perspective on the free will problem by allowing for a contextualist solution to it; one according to which the truth conditions of freedom ascriptions vary across contexts of utterance.

For centuries philosophers have been arguing back and forth about free will. Incompatibilists appeal to the strong intuition that nothing counts as a free action if it has been brought about via a deterministic causal chain which does not ultimately go back to the agent. Compatibilists emphasize a much more moderate conception of what it is for an action to be free – on their view, freedom is very well compatible with determinism. Both competing sides claim that their conception of free will meets the commonsensical, or at least the one and only relevant understanding of that notion; each side insists on capturing just the right kind of intuitions about freedom of the will in their account.

Which moral should be drawn? It has been argued that the debate has reached an impasse (van Inwagen 2000) or that our ordinary concept of freedom is incoherent (Double 1991; Jackson 1998). Recently, John Hawthorne (2001) has made a more constructive suggestion: freedom statements, statements in which freedom of the will is ascribed to or denied of some subject, might be context sensitive in the sense that they have different truth conditions across different contexts of use.²

Concerning the conflict between the compatibilist and the incompatibilist, this suggests that freedom statements have different truth conditions across ordinary contexts and contexts in which determinism comes into focus. And while the ordinary truth conditions for freedom ascriptions are often easily met, they are too demanding to be met in contexts in which we focus on determinism. This allows for a moderate compatibilism, since the truth of determinism does not inflict the truth of our ordinary freedom ascriptions. Let's dub this view "freedom contextualism".³

If freedom contextualism is on the right track, then any account which ignores the context sensitivity of freedom statements inevitably falls short of accounting for a pertinent class of intuitions about freedom. If successful, the view therefore explains the dead-end situation between compatibilists and incompatibilists and offers a semantics of freedom statements which does justice to the most pertinent intuitions in favor of each of the competing views. In doing so, freedom contextualism provides a completely new perspective on one of the knottiest imbroglios philosophy has to offer.⁴

Unfortunately, the view does not seem particularly compelling as Hawthorne presents it.⁵ Let's look at the view. According to Hawthorne's proposal,

> REA. S does x freely only if S's action is free from causal explainers beyond S's control – Psst! – except for those explainers that we are properly ignoring. (ibid.)

Let's call this view "the relevant explainers account". A causal explainer, according to Hawthorne, is "simply a state of affairs which provides an adequate causal explanation of an action" (ibid.). The truth of a freedom statement is then construed as depending on both the "causal influences upon action" and "the context of attention" (ibid.). This yields the desired freedom contextualist solution to the compatibility problem of freedom and determinism, because

> [w]hen ordinary speaker utter English claims of the form 'S did x freely' (and their synonyms), they frequently speak the truth. When we approach a God's eye perspective on the causal nexus, we become ever bolder when claiming that S does x freely – for as we approach that perspective, the causes that we are properly ignoring diminishes. In the con-

---

2 Contextualism is very influential in epistemology. See for instance DeRose (2009).
3 Freedom contextualism has also been suggested by Rieber (2006). I'll focus on Hawthorne's version of the view in what follows.
4 But see Schulte (2014) for an interesting objection to freedom contextualism.
5 Note that Hawthorne only toys with the view.

text of ordinary life, by contrast, the causes that are properly ignored are much greater in number and thus it is much easier for ordinary freedom ascriptions to be true. (ibid.: 69)

Hawthorne's presentation of REA remains pretty much in the abstract. He does not give examples for types of causal explainers we may properly ignore in ordinary life, but not in the face of the consideration of determinism. One natural way to flesh out what he may have in mind is this, though: in many ordinary contexts, we may ignore all but those causal explainers that are internal to the agent; all we need to care about are the agent's motivational states, for instance. Freedom ascriptions made in contexts of that kind are true because the motivational states themselves are within the agent's control.

In contexts in which determinism is considered, in contrast, we also have to take the innumerable causal explainers prior to the agent's willing into account. Hence, the condition stated in REA fails to be met in such contexts: the state of the world at some time before the agent's birth together with the laws of nature, say, is not in the agent's control. Hence, our freedom ascriptions turn out false in the face of determinism.

So much for Hawthorne's line of argument. Unfortunately, it is not very convincing. There are many problems, of which I want to highlight just one.[6] REA works with the notion of control. But in doing so, it obviously begs the question against the incompatibilist. Incompatibilists will typically reject the idea that we have control over any of our motivational states (or in fact about anything whatsoever), if these motivational states (or whatever other causal explainer we are focusing on) are in turn determined by events over which we lack any kind of control. In fact, this is but one way of phrasing what the compatibilism debate is about. As Peter Schulte rightly points out,

> [t]he term "control" is (...) vague and ambiguous. We speak of people controlling their movements or their facial expression, but we also say that the autonomic nervous system controls respiration, or that a thermostat controls the room temperature. It is clear, therefore, that in some ordinary sense of "control", determined agents can exercise control over their actions, but there is presumably also an ordinary sense of this term in which control and determinism are incompatible. (Schulte 2014: 673)

REA fails to factor this dimension of the debate in; it simply *presupposes* that the compatibilist gets the matter about control right. This is a very serious problem. In presupposing that compatibilism wins over incompatibilism from the outset,

---

6 I point out a variety of further problems in Jaster (forthcoming).

REA undermines its own motivation – it makes a contextualist framework superfluous as a solution for the compatibilism issue.

Is Freedom Contextualism a nonstarter, then? Let's not give up too soon. Here is a more plausible freedom contextualist analysis:

> PAP. S does x freely only if S can act otherwise than x.

But wait, you may think. How is this a contextualist analysis of freedom statements? Isn't this just a version of the well-known principle of alternate possibilities?[7] Correct. PAP is nothing other than that. What makes this a *contextualist* analysis is the insight that ability statements quite generally, and thus statements of the form "S can do otherwise" *are* in fact context sensitive. Let me explain.

The condition on freedom formulated in PAP crucially contains an ability ascription: "S can do otherwise". Since acting otherwise will by definition be an action, the ability that is ascribed is an agentive ability. Now let's plug in what we know about ability statements: they are context sensitive. According to the simplified version of the success view of agentive abilities,

> SUCCESS$_{AA}$. an agent has an agentive ability to ϕ if and only if the agent ϕ's in a sufficient proportion of the relevant possible situations in which the agent intends to ϕ, where it varies extensively across ascriber contexts which situations are relevant.[8]

We are now in a position to formulate our freedom contextualist analysis more explicitly. For we can now draw on PAP and the success view to formulate the new view that PAP, rightly understood, translates into what I will call Alternate Possibilities Contextualism, or APC, for short:

> APC. S does x freely only if S does something other than x in a sufficient proportion of the relevant possible situations in which S intends to do something other than x, where it will vary across contexts which situations are relevant.[9]

---

7 Originally, the principle was formulated in terms of responsibility instead of freedom (Frankfurt 1969) and in past tense ("could have done otherwise").
8 This is a simplification in so far as it neglects that the action type that the agent intends to perform need not be (*de re* or *de dicto*) identical to ϕ-ing, as long as ϕ-ing vis-à-vis that intention counts as a success ($\rightarrow$ 4.8).
9 Note again that "intending to do something other than x" can be given the *de dicto* or *de re* reading here.

APC has obvious merits. On the one hand, it derives the context sensitivity of "free" from an independently motivated paradigm about freedom: freedom requires alternate possibilities.[10] On the other hand, it is based on the success view of ability, which is motivated independently, in virtue of its merits as a view of abilities quite generally. It is hard to think of ways in which a freedom contextualist analysis could be better motivated than that.

Note as well that, in contrast to Hawthorne's REA, APC does not beg the question against the incompatibilist. Quite the contrary. The problem about the compatibility of freedom and determinism is traditionally taken to arise primarily against the background of PAP (van Inwagen 1983, ch. 1). PAP is therefore one of the crucial premises in the incompatibilist's line of reasoning. APC starts off from this very premise. It is therefore fully in line with the incompatibilist's set of initial assumptions.

How does APC solve the puzzle about free will? The puzzle about free will, recall, is that we are inclined to ascribe freedom on a regular basis. But once determinism comes into focus, our entitlement to such ascriptions is jeopardized. On the general freedom contextualist account, the puzzle can be resolved once we notice that our ordinary claims to "freedom" and our denials of "freedom" in the face of determinism need not be contradictory. On the APC framework, that variance in truth conditions of freedom claims traces back to a variance in truth conditions of the underlying "S can act otherwise" statements. So, let's see whether the truth conditions of "can" statements vary in the right way to explain the postulated variances in our statements about freedom. What needs to be shown is that the modal base of an "S can act otherwise" statement is different when determinism is considered than in ordinary contexts.

This seems indeed plausible. When determinism is in focus, the totality of all actual facts up to the moment in which the action is performed (including facts about the natural laws) determine the modal base. In other words, the modal base will comprise only situations in which the totality of facts is preserved. In contexts of this kind, "S can act otherwise" statements turn out false. No one can perform any action in view of the fact that she is determined not to perform it. If the modal base comprises only situations in which the totality of facts up to the moment of the action is preserved, then it does not comprise any in which the agent intends to act otherwise or acts otherwise. In fact, the modal base contains only the actual situation itself in that case.

---

10 Of course, the principle is not uncontroversial. See Widerker & McKenna (2003) for important contributions to the controversy.

In ordinary contexts, this is very different. It will not be required that the totality of the actual facts be preserved in that world or that some prior state of the world and the laws necessitate some act. That is just what makes the context ordinary, as opposed to one in which determinism is considered. Which facts *do* go into the modal base in ordinary contexts? – That depends. Ordinary contexts have but one thing in common: determinism is not considered. And as we know from the consideration of various ordinary context ascriptions of abilities throughout the previous chapters of this book, there is not one specific set of facts that is relevant in all ordinary contexts. Sometimes, we are interested in what someone can do in view of the program in her brain, sometimes, in view of her broken limb, sometimes, in view of her dizziness, hoarseness, phobia, and so forth.

If this is true then the truth conditions of "S can act otherwise" statements will vary across ordinary contexts to the very same extent that they vary across ordinary contexts on the one hand and contexts in which determinism is considered on the other. That is not problematic, though. As long as "S can act otherwise" statements often turn out true in ordinary contexts, we can solve the puzzle about freedom just fine. And they *will* often come out true! In view of my muscular constitution, I can act otherwise than shying away from the spider, say, even though I cannot act otherwise in view of my phobia. The success view accounts for both. I act otherwise than shying away from the spider in a sufficient proportion of the relevant possible intention situations in which my muscular constitution is held fixed (and my phobia varied), but it is not the case that I act otherwise in a sufficient proportion of the relevant possible intention situations in which the phobia is held fixed.[11]

Let me wrap up. I argued that Hawthorne's freedom contextualism is not particularly appealing, because it begs the question against incompatibilists about freedom. I then proposed a different freedom contextualist view. Alternate Possibilities Contextualism (APC) starts off from a version of the Principle of Alternate Possibilities and combines that principle with the success view of ability. For an agent to be free, I argued, it has to be true that the agent acts otherwise in a sufficient proportion of the relevant possible situations in which the agent intends to act otherwise, where it will vary by context which situations are relevant. The view is well-motivated, does not beg the question against the incompatibilist, and explains the context-shifts posited by the freedom contextualist.

---

11 In Jaster (forthcoming), I argue that the truth values of freedom ascriptions vary in accordance to these variances in "S can act otherwise" statements. If this is true, then the truth conditions of freedom statements vary across ordinary contexts to the same amount as they vary across ordinary contexts and contexts in which determinism is considered.

We can conclude that freedom contextualism, if plausible at all, is quite plausibly spelled out in terms of APC.

The question, of course, is how plausible freedom contextualism as whole turns out to be. Among other things, the answer to this question hinges crucially on whether or not a contextualist semantics of freedom ascriptions can be bolstered on the basis of our ordinary practice of freedom ascriptions.[12] I have not addressed that question and I wish to remain neutral on it.[13] My main purpose here was to show how the success view of ability can be put to work when it comes to actual philosophical problems. And as I hope to have shown, the view provides an interesting toolbox, which allows for a fresh assessment of some well-established problems.

---

12 Epistemic contextualists have gone to great lengths to establish such ordinary language evidence for contextualism about knowledge. See for instance DeRose (2005).
13 But see again Jaster (forthcoming) for a discussion.

Resumé and an open question

Let me give you a very brief upshot of the main findings of this book. I have defended what I called "the success view of ability". On that view (→ 5.3),

> SUCCESS$_{ABILITY}$. an agent S has an ability to φ if and only if S φ's in a sufficiently high weighted proportion of the relevant possible S-trigger situations.

An S-trigger situation, on that view, is a situation in which some trigger is present in response to which φ-ing counts as a success.

Abilities and dispositions, I suggested (→ 5.4), differ along the success dimension. While it seems plausible that an object has a disposition to φ if and only if the object φ's in a sufficiently high proportion of the relevant possible situations in which some trigger for φ-ing is present, that trigger will have to be an S-trigger in order for that disposition to qualify as an ability.

The most paradigmatic cases of abilities are what I called "agentive abilities" – abilities to perform actions. The S-trigger that matters in the case of such abilities, I argued (→ 4.2, 4.8, 4.9), is the (*de dicto* or *de re*) intention to φ, or – in exceptional cases – a different intention entirely, where that intention, too, will have to be an S-trigger. Thus, according to the success view of agentive abilities (→ 4.10),

> AGENTIVE ABILITIES. an agent S has an agentive ability to φ if and only if S φ's in a sufficiently high proportion of the relevant possible situations in which S intends to ψ, where ψ will typically be *de dicto* or *de re* identical to φ, and in all other cases be such that φ-ing in response to the intention to ψ-ing counts as a success.

The success view is a hybrid of conditional analyses (→ 2, 6.2) and possibilism (→ 3). On the one hand, it analyzes agentive abilities in terms of a modal tie between the agent's motivation and the corresponding performances, which is a core feature of conditional analyses. On the other hand, it establishes that modal tie by means of restrictions on the possible situations, which is a core feature of possibilism.

The view, I argued (→ 4.3 – 4.7), is very powerful. It accounts for masks and cases of impeded intentions; it provides an account of degrees and the corresponding kind of context sensitivity of ability ascriptions; it provides an understanding of general as well as specific abilities and the relationship between the two; finally, it provides an understanding of agentive and non-agentive abilities and how they relate. The view therefore meets all of the adequacy conditions for a comprehensive view of abilities that were laid out in chapter 1.

Some topics are underdeveloped in this book and I want to take the time to mark out one of those topics very clearly in the end. In part, this is for reasons of academic virtuousness, in part, however, it is also because I take this open flank to set a highly interesting agenda for future research.

I have said nothing about the ability to intend. When I talked about agentive abilities, I explained that the success view of such abilities does not run into the problem of the impeded intention (→ 4.4). The reason it does not run into that problem is that it entails what I called the existential requirement: there has to be a relevant intention situation to begin with, in order for there to be a sufficiently high proportion of performance cases *among* the relevant intention situations. And this requirement, I laid out, fails to be met in cases in which an agent cannot intend to φ in the first place. That was all we needed to shield the success view from the problem of the impeded intention. So far, so good, then.

What I haven't talked about, however, is the ability to intend to φ as such. Quite plausibly, meeting the existential requirement is not *sufficient* for the having of the ability to intend to φ, after all. The reason is simple. Agents can have the ability to form a certain intention to different degrees. The existential requirement is therefore just as ill-suited as a sufficient condition for the having of the ability to intend to φ as possibilism was for the ability to φ itself (→ 3.5).

That the ability to intend, too, comes in degrees may not be obvious right away, because we usually do not grade the ability to intend. We do not say things like "She is better at intending to cook" or "He is the better intender when it comes to cooking". However, and this is important, the ability to intend can be diminished to varying degrees. And this shows that the existential requirement is only a necessary, but not a sufficient condition for the having of the ability to intend. Let me explain.

So far (→ 2.3, 4.4), we have only ever thought about cases of impeded intentions as an all or nothing matter. The coma patient lacks the ability to raise her arm, while I have it, because she cannot intend to raise her arm, while I can. The brainwashed follower of a cult cannot leave the cult, while the non-brainwashed person can, because the brainwashed follower cannot form the intention to leave, while the non-brainwashed person can. A phobic, to take another famous example, cannot touch a spider, while normal people can, because the phobic cannot form the intention to touch it, while normal people can. In each case, the ability to intend to φ is fully impaired in one case, and fully intact in the other.

But now think of what we can call a "mild phobic". A phobic, that is, who sometimes, in certain therapy sessions, manages to form the intention to touch a spider. And then *does* touch a spider! Mind you: her motor abilities are not impaired in any way. Compare her ability to touch a spider with the ability of what we can call a "full-fledged phobic". A phobic, that is, who cannot form the intention to touch spiders at all and therefore never manages to touch a spider. It seems clear that one of them is better at touching spiders than the other one. And where does this difference in the degree of the ability to touch spiders stem from? Quite obviously, it stems from a difference in the degree to which they form the intention to touch a spider. That, after all, is the only difference between the two agents.

The problem with the existential requirement is that it cannot capture differences in degrees of the ability to form intentions. As we know from our discussion of possibilism (→ 3.5), the existential quantifier is ill-suited to do that job. All this is unsurprising. The ability to intend, just like any other ability, will have to be analyzed in terms of the success view, of course. It will be a matter of proportions among properly restricted situations.

Does this lead into a regress? Does the ability to touch spiders require an ability to intend to touch spiders and does that ability, in turn, require an ability to intend to intend to touch spiders, and so forth? It does not. Recall that the success view of agentive abilities is a special instance of the success view of abilities *tout court*. And that stops the regress.

The ability to touch spiders is an agentive ability. It requires, among other things, that the agent have the ability to intend to touch spiders.[1] And for that ability, the existential requirement postulates a necessary condition (→ 4.4). To have the ability to touch spiders, there have to be relevant possible situations in which one forms the intention to touch spiders.

The ability to *intend* to touch spiders, however, is not itself an agentive ability. Intending is not an action. That is why I formulated the success view of agentive abilities in terms of intentions in the first place, and not in terms of decisions or other mental acts that seem to be actions themselves (→ 4.2).

The ability to intend will thus have to be analyzed along the lines of the success view of non-agentive abilities. Having an ability to intend to φ is for the agent to intend to φ in a sufficient proportion of the relevant possible situations in which some S-trigger for intending to φ is present. This brings us to the question I have not answered in this book. Which S-trigger matters in the case of intentions? How can we analyze the ability to form an intention? This is a question that has not been answered and one that will have to remain open for the time being.

---

[1] Or form some other suitable intention, which is properly modally linked to touching spiders (→ 4.8). For reasons of simplicity, I neglect this subtlety here.

I sympathize with the idea that having an ability to intend to φ is a matter of intending to φ in a sufficient proportion of the relevant possible situations in which there is an overriding reason to intend to φ. But this condition is not easily spelled out. Are the reasons which matter motivational or normative reasons? Or are both potential S-triggers? There is obviously some more work to do here, and I cannot say that I have made up my mind about these things.

I'll leave that for another day. Or year.

References

Adams, F., 1994, "Trying, Desire, and Desiring to Try", *Canadian Journal of Philosophy* 24(4): 613–626.
Adams, R., 1974, "Theories of Actuality", *Noûs* 8: 211–231; reprinted in Loux (ed.), 1979, *The Possible and the Actual*. Ithaca, NY: Cornell University Press, 190–209.
Aristotle, 1999, *Metaphysics*. Joe Sachs (trans.), Santa Fe, NM: Green Lion Press.
Armstrong, D.M., Martin, C.B. & Place, U.T., 1996, *Dispositions: A Debate*. London: Routledge.
Aune, B., 1963, "Abilities, modalities, and free will", *Philosophy and Phenomenological Research* 23: 397–413.
Austin, J.L., 1956, "Ifs and Cans," *Proceedings of The British Academy* 42: 107–132.
Ayer, A. J., 1946, "Freedom and Necessity", *Polemic* 5: 38–51; reprinted in Ayer, 1954, *Ayer: Philosophical Essays*, London: Greenwood Press, 271–284.
Berofsky, B. (ed.), 1966, *Free Will and Determinism*. New York: Harper & Row.
Berofsky, B., 2002, "Ifs, cans, and free will: The issues"; in Kane, R.H. (ed.), *The Oxford Handbook of Free Will*. Oxford: Oxford University Press, 181–201.
Bigelow, J. C., 1976, "Possible Worlds Foundations for Probability", *Journal of Philosophical Logic* 5(3): 299–320
Bird, A., 1998, "Dispositions and Antidotes", *The Philosophical Quarterly* 48: 227–234.
Bird, A., 2005, "The dispositionalist conception of laws", *Foundations of Science* 10(4): 353–370.
Bird, A., 2007, *Nature's Metaphysics: Laws and Properties*. Oxford: Oxford University Press.
Chisholm, R. M., 1976, *Person and Object: A Metaphysical Study*. La Salle, IL: Open Court.
Chisholm, R. M., 1966, "Freedom and Action"; in Lehrer, K. (ed.), *Freedom and Determinism*. New York: Random House: 11–44; reprinted in Brand, M. (ed.), 1970, *The Nature of Human Action*. Glenview, IL: Scott Foresman & Co., 283–292.
Choi, S., 2006, "The Simple vs. Reformed Conditional Analysis of Dispositions", *Synthese* 148: 369–379.
Choi, S., 2008, "Dispositional Properties and Counterfactual Conditionals", *Mind* 117: 795–841.
Clarke, R., 2009, "Dispositions, Abilities to Act, and Free Will: The New Dispositionalism", *Mind* 118: 323–351.
DeRose, K., 2005, "The Ordinary Language Basis for Contextualism and the New Invariantism", *The Philosophical Quarterly* 55(219): 172–198.
DeRose, K., 2009, *The Case for Contextualism*. Oxford: Clarendon Press.
Double, R., 1991, *The Non-Reality of Free Will*. Oxford: Oxford University Press.
Evans, G., 1982. *The Varieties of Reference*. Oxford: Oxford University Press.
Fara, M., 2005, "Dispositions and Habituals", *Noûs* 39: 43–82.
Fara, M., 2008, "Masked Abilities and Compatibilism", *Mind* 117: 843–865.
Fine, K., 1977, "Postscript"; in Prior, A.N. (ed.), *Worlds, Times, and Selves*. Amherst: University of Massachusetts Press, 116–161.
Frankfurt, H., 1969, "Alternate Possibilities and Moral Responsibility", *Journal of Philosophy* 66: 829–839.
Geach, P., 1957, *Mental Acts, Their Content and Their Objects*. London, New York: Routledge & Kegan.

Goldman, A., 1979, "What Is Justified Belief?"; in Pappas, G. S. (ed.), *Justification and Knowledge*, Dordrecht: Reidel, 1–25; reprinted in A. I. Goldman, 2012, *Reliabilism and Contemporary Epistemology*. New York: Oxford University Press, 29–49.

Goodman, N., 1954, *Fact, Fiction and Forecast*. Cambridge, MA: Harvard University Press.

Greco, J., 1999, "Agent Reliabilism"; in J. Tomberlin (ed.), *Philosophical Perspectives* 13: Epistemology. Atascadero: Ridgeview, 273–296.

Greco, J., 2003, "Knowledge as Credit for True Belief"; in M. DePaul & L. Zagzebski (eds.), *Intellectual Virtue: Perspectives from Ethics and Epistemology*, Oxford: Oxford University Press, 111–134.

Greco, J., 2009, "Knowledge and Success From Ability", *Philosophical Studies* 142: 17–26.

Greco, J., 2010, *Achieving Knowledge: A Virtue-Theoretic Account of Epistemic Normativity*. Cambridge: Cambridge University Press.

Greco, J., & Turri, J., 2015, "Virtue Epistemology", *The Stanford Encyclopedia of Philosophy* (Fall 2015 Edition), Edward N. Zalta (ed.), URL = <http://plato.stanford.edu/archives/fall2015/entries/epistemology-virtue/>.

Gundersen, L., 2002, "In Defence of the Conditional Account of Dispositions", *Synthese* 130: 389–411.

Harré, R. & Madden, E. H., 1975, *Causal Powers: A Theory of Natural Necessity*. Oxford: Basil Blackwell.

Harré, R., 1970, "Powers", *British Journal for the Philosophy of Science* 21: 81–101.

Hawthorne, J., 2001, "Freedom in Context", *Philosophical Studies* 104: 63–79.

Heller, M., 1999, "The Proper Role for Contextualism in an Anti-Luck Epistemology", *Philosophical Perspectives* 13: 115–129.

Honoré, A. M., 1964. "Can and Can't," *Mind* 73: 463–479.

Horgan, T., 1979, "Could, Possible Worlds, and Moral Responsibility", *Southern Journal of Philosophy* 17(3): 345–358.

Hume, D., 1748 (1999), *Enquiry concerning Human Understanding*. T.L. Beauchamp (ed.), Oxford/New York: Oxford University Press.

Jackson, F., 1998, *From Metaphysics to Ethics*. Oxford: Clarendon Press.

Jaster, R., forthcoming, "Contextualizing Free Will", Zeitschrift für philosophische Forschung.

Johnston, M., 1992, "How to Speak of the Colors", *Philosophical Studies* 68: 221–263.

Keil, G., 2007, *Willensfreiheit*. Berlin: de Gruyter.

Kennedy, C., 2007, "Vagueness and grammar: the semantics of relative and absolute gradable adjectives", *Linguistics and Philosophy* 30(1): 1–45.

Kenny, A., 1976, "Human Abilities and Dynamic Modalities"; in Manninen & Tuomela (eds.), *Essays on Explanation and Understanding*, Dordrecht: D. Reidel Publishing Company, 209–232.

Kratzer, A., 1977, "What 'Must' and 'Can' Must and Can Mean", *Linguistics and Philosophy* 1: 337–355.

Kratzer, A., 1981. "The Notional Category of Modality"; in Eikmeyer and Rieser (eds.), *Words, Worlds, and Contexts: New Approaches in Word Semantics*. Berlin: de Gruyter, 38–74.

Kratzer, A., 1991, "Modality"; in von Stechow, A. & Wunderlich, D. (eds.), *Semantics: An International Handbook of Contemporary Research*. Berlin: de Gruyter, 639–650.

Lehrer, K., 1968, "Cans without Ifs", *Analysis* 29: 29–32.

Lehrer, K., 1976, "'Can' in Theory and Practice: A Possible World Analysis"; in Brand and Walton (eds.), *Action Theory*. Dordrecht: D. Reidel, 241–270.

Lewis, D., 1973, *Counterfactuals*. Cambridge, MA: Harvard University Press; reissued London: Blackwell, 2001.
Lewis, D., 1976, "The Paradoxes of Time Travel", *American Philosophical Quarterly* 13: 145–152.
Lewis, D., 1981, "Ordering Semantics and Premise Semantics for Counterfactuals", *Journal of Philosophical Logic* 10: 217–234.
Lewis, D., 1986, *On the Plurality of Worlds*. London: Blackwell.
Lewis, D., 1996, "Elusive Knowledge", *Australasian Journal of Philosophy* 74(4): 549–567.
Lewis, D., 1999. "What Experience Teaches"; in Lewis, D. (ed.), *Papers in Metaphysics and Epistemology*. Cambridge: Cambridge University Press, 262–290.
Lewis, D., 1997, "Finkish Dispositions", *The Philosophical Quarterly* 47: 143–158.
Locke, J., 1690, *An Essay Concerning Human Understanding*. P. H. Nidditch (ed.), Oxford: Oxford University Press, 1975.
Löwenstein, D., 2013, "Why Know-how and Propositional Knowledge Are Mutually Irreducible"; in Hoeltje, Spitzley & Spohn (eds.), *Was dürfen wir glauben? Was sollen wir tun? – Sektionsbeiträge des achten internationalen Kongresses der Gesellschaft für Analytische Philosophie e.V.*. Universität Duisburg-Essen: DuEPublico, 365–371.
Löwenstein, D., 2017, *The Concept of Competence. A Rylean Responsibilist Account of Know-how*. Frankfurt am Main: Vittorio Klostermann.
Mackie, J. L., 1973, *Truth, Probability and Paradox*. Oxford: Oxford University Press.
Maier, J., 2013, "The Agentive Modalities", *Philosophy and Phenomenological Research* 87: 1–22.
Maier, J., 2014, "Abilities", *The Stanford Encyclopedia of Philosophy* (Fall 2014 Edition), Edward N. Zalta (ed.), URL = <http://plato.stanford.edu/archives/fall2014/entries/abilities/>.
Makin, S. (trans.), 2006. *Metaphysics, Book Θ*. Oxford: Oxford University Press.
Manley, D. & Wasserman, R., 2008, "On Linking Dispositions and Conditionals", *Mind* 117: 59–84.
Mayr, E., 2011, *Understanding Human Agency*. Oxford: Oxford University Press.
McKitrick, J., 2003, "A Case for Extrinsic Dispositions", *Australasian Journal of Philosophy* 81: 155–174.
Mele, A., 2002. "Agents' Abilities," *Noûs* 37: 447–470.
Menzel, C., 2016, "Possible Worlds", *The Stanford Encyclopedia of Philosophy* (Spring 2016 Edition), Edward N. Zalta (ed.), forthcoming URL = <http://plato.stanford.edu/archives/spr2016/entries/possible-worlds/>.
Menzies, P., 1998, "Possibility and conceivability: A response-dependent account of their connections"; in Casati & Tappolet (eds.), *European Review of Philosophy*, 3 "Response-Dependence". CSLI Press.
Menzies, P., 2014, "Counterfactual Theories of Causation", *The Stanford Encyclopedia of Philosophy* (Spring 2014 Edition), Edward N. Zalta (ed.), URL = <http://plato.stanford.edu/archives/spr2014/entries/causation-counterfactual/>. Last retrieved: Jan 4, 2020
Millikan, R. G., 1984, *Language, Thought and Other Biological Categories*. Cambridge, MA: MIT Press.
Millikan, R. G., 2000, *On Clear and Confused Ideas*. Cambridge: Cambridge University Press.
Molnar, G., 1999, "Are Dispositions Reducible?", *The Philosophical Quarterly* 49: 1–17.

Moore, G. E., 1912, *Ethics*. London: Williams & Norgate.
Morriss, P., 2012, *Power*. Manchester/New York: Manchester University Press.
Mumford, S., 1998, *Dispositions*. Oxford: Oxford University Press.
Mumford, S., 2004, *Laws in Nature*. New York: Routledge.
Mumford, S., & Anjum, R. L., 2011, *Getting Causes from Powers*. Oxford: Oxford University Press.
Nozick, R., 1981, *Philosophical Explanations*. Oxford: Oxford University Press.
Palmer, F. R., 1977, "Modals and Actuality", *Journal of Linguistics* 13(1): 1–23.
Peacocke, C., 1999. *Being Known*. Oxford: Oxford University Press.
Plantinga, A., 1974, *The Nature of Necessity*. Oxford: Oxford University Press.
Plantinga, A., 1976, "Actualism and Possible Worlds", *Theoria* 42: 139–160.
Prior, A. N., 1977, *Worlds, Times, and Selves*. Amherst: University of Massachusetts Press.
Prior, E., Pargetter, R. & Jackson, F., 1982, "Three Theses about Dispositions", *American Philosophical Quarterly* 19: 251–257.
Quine, W. V. O., 1956, "Quantifiers and propositional attitudes", *Journal of Philosophy* 53: 177–187.
Quine, W. V. O., 1960, *Word and Object*, Cambridge, MA: MIT Press.
Rieber, S., 2006, "Free Will and Contextualism", *Philosophical Studies* 129: 223–252.
Ryle, G., 1949, *The Concept of Mind*. London: Hutchinson.
Schlick, M., 1939, "When is a Man Responsible?", in Rynin, D. (trans.), *Problems of Ethics*, 143–146. Reprinted in Berofsky (ed.), 1966, *Free Will and Determinism*. New York: Harper & Row.
Schulte, P., 2014, "Beyond Verbal Disputes: The Compatibilism Debate Revisited", *Erkenntnis* 79: 669–685.
Shoemaker, S., 1980, "Causality and Properties", in van Inwagen, P. (ed.), *Time and Cause: Essays Presented to Richard Taylor*. Dordrecht: Reidel, 109–135.
Simpson, J. & Weiner, E. (eds.), 1989, *The Oxford English Dictionary (2nd edition)*. Oxford: Oxford University Press.
Smith, M., 2003, "Rational Capacities"; in Stroud & Tappolet (eds.), *Weakness of Will and Varities of Practical Irrationality*. Oxford: Oxford University Press, 17–38.
Sosa, E., 1991, *Knowledge in Perspective*. Cambridge: Cambridge University Press.
Sosa, E., 2007, *Apt Belief and Reflective Knowledge, Volume 1: A Virtue Epistemology*. Oxford: Oxford University Press.
Sosa, E., 2011, *Knowing Full Well*. Princeton: Princeton University Press.
Sosa, E., 2015, *Judgment and Agency*. Oxford: Oxford University Press.
Stalnaker, R., 1968, "A Theory of Conditionals", *American Philosophical Quarterly* Monograph 2: 98–112.
van Inwagen, P., 1978, "Ability and Responsibility", *The Philosophical Review* 2: 201–224.
van Inwagen, P., 1983, *An Essay on Free Will*. Oxford: Oxford University Press.
van Inwagen, P., 2000, "Free Will Remains a Mystery"; in Tomberlin, J. (ed.), *Philosophical Perspectives, 14: Action and Freedom*, Hoboken, NJ: Blackwell Publishing, 1–19.
van Inwagen, P., 2008, "How to think about the problem of free will", *Journal of Ethics* 12: 327–341.
Vetter, B., ms., "Understanding Abilities. An opinionated survey".
Vetter, B., 2011, "On Linking Dispositions and which Conditionals?", *Mind* 120: 1173–1189.
Vetter, B., 2015, *Potentiality: From Dispositions to Modality*. Oxford: Oxford University Press.

Vihvelin, K., 2004. "Free Will Demystified: A Dispositionalist Account," *Philosophical Topics* 32: 427–450.
Vihvelin, K., 2013, *Causes, Laws, and Free Will: Why Determinism Doesn't Matter*. Oxford: Oxford University Press.
von Fintel, K., 2006, "Modality and Language"; in Borchert, D. (ed.), *Encyclopedia of Philosophy*. New York: Macmillan Reference: 20–27.
von Wright, G. H., 1951, *An Essay in Modal Logic*. Amsterdam: North-Holland Pub. Co.
Vranas, P. B. M., 2009, "Can I kill my younger self? Time travel and the retrosuicide paradox", *Pacific Philosophical Quarterly* 90: 520–534.
Whittle, A., 2008, "A Functionalist Theory of Properties", *Philosophy and Phenomenological Research* 77: 59–82.
Whittle, A., 2010, "Dispositional Abilities," *Philosophers' Imprint*, 10(12): 1–23.
Widerker, D., 1995, "Libertarianism and Frankfurt's Attack on the Principle of Alternative Possibilities", *The Philosophical Review* 104: 247–261.
Widerker, D., & McKenna, M. (eds.), 2003, *Moral Responsibility and Alternative Possibilites*, Adlershot: Ashgate Publishing.
Wilson, G. & Shpall, S., "Action", *The Stanford Encyclopedia of Philosophy* (Summer 2012 Edition), Edward N. Zalta (ed.), URL = <http://plato.stanford.edu/archives/sum2012/entries/action/>.
Yablo, S., 1993, "Is Conceivability a Guide to Possibility?", *Philosophy and Phenomenological Research* 53(1): 1–42.
Zagzebski, L., 1996, *Virtues of the Mind*. Cambridge: Cambridge University Press.
Zagzebski, L., 2009, *On Epistemology*. Belmont, CA: Wadsworth.

Index of Names

Adams, F.   41, 102
Adams, R.   41, 102
Aristotle   18, 166
Armstrong, D.M.   8
Aune, B.   46–48
Austin, J.L.   122, 125
Ayer, A. J.,   38

Berofsky, B.   19 f., 115 f., 119–121
Bigelow, J. C.   175
Bird, A.   32, 47 f.

Chisholm, R. M.   49
Choi, S.   47, 165
Clarke, R.   54, 163 f., 168

DeRose, K.   219, 224
Double, R.   218

Evans, G.   1

Fara, M.   2 f., 32, 48, 54, 125, 163–165, 167, 180 f.
Fine, K.   41
Frankfurt, H.   2, 214, 221

Geach, P.   86, 117
Goldman, A.   184
Goodman, N.   46
Greco, J.   1, 96, 174, 182–188, 212, 214
Gundersen, L.   47

Harré, R.   32
Hawthorne, J.   218–220, 223
Heller, M.   28, 41
Honoré., A. M.   21 f., 114, 117
Horgan, T.   5, 63, 69 f.
Hume, D.   4, 38

Jackson, F.   8, 218
Jaster, R.   220, 223 f.
Johnston, M.   44

Keil, G.   167
Kennedy, C.   26
Kenny, A.   72, 85–88, 120, 137 f., 140, 152
Kratzer, A.   2, 5, 35, 63–71, 73, 79, 83, 85, 98, 114, 128, 174, 177, 199, 207

Lehrer, K.   4 f., 49 f., 63
Lewis, David   1, 5, 8 f., 15, 31, 40–42, 47, 63, 68–70, 141, 195, 215–218
Locke, J.   166
Löwenstein, D.   6, 29, 141, 154

Mackie, J. L.   8
Madden, E. H.   32
Maier, J.   18 f., 21 f., 24 f., 28, 103, 114, 123, 174, 185, 205–208, 210–213
Manley, R.   4, 23, 33, 49, 94, 103 f., 106, 174 f., 181 f., 184, 188–192, 200, 212
Mayr, E.   1, 59, 166, 214
McKitrick, J.   167, 189
Mele, A.   20, 86, 125
Menzel, C.   41
Menzies, P.   1, 41 f., 214
Millikan, R. G.   1, 25, 34, 165, 170, 214
Molnar, G.   141
Moore, G. E.   2–5, 32, 38, 49, 52, 102, 162, 180
Morriss, P.   16
Mumford, S.   32, 47, 165

Nozick, R.   43

Palmer, F. R.   21
Peacocke, C.   112
Plantinga, A.   41
Prior, A. N.   8

Quine, W. V. O.   46, 149

Rieber, S.   219
Ryle, G.   2, 32, 46, 162, 180

Schlick, M.   38

Schulte, P.   219 f.
Shpall, S.   101, 150
Simpson, J.   95
Smith, M.   2 f., 32, 163, 167, 180
Sosa, E.   1, 4, 25, 162, 174, 183 f., 188, 191–194, 196, 212, 214
Stalnaker, R.   31, 41 f.

van Inwagen, P.   3, 29, 49, 78, 214 f., 218, 222
Vetter, B.   8, 20, 32, 59, 65 f., 84, 115 f., 143, 165, 169, 174, 182, 185, 187, 189, 191, 199–201, 203, 213
Vihvelin, K.   2 f., 19, 21, 32, 54, 163, 167, 180 f., 184, 188, 192–194, 196, 212

von Wright, G. H.   5, 63, 125
Vranas, P. B. M.   217

Wasserman, D.   4, 23, 33, 49, 94, 103 f., 106, 174 f., 181 f., 184, 188–192, 200, 212
Weiner, E.   95
Whittle, A.   19–22, 32, 47, 49, 114, 121, 123, 125, 164
Widerker, D.   3, 222
Wilson, G.   101, 150

Yablo, S.   1, 214

Zagzebski, L.   183

Index of subjects

Ability to do otherwise   2–5, 167 f., 213 f., 221–223
Ability to Intend   52, 226–228
Achievement   25 f., 56, 81 f., 103–108, 158, 203
Actions   1, 4–6, 10, 13 f., 21, 24, 28–31, 34 f., 38, 49, 54, 58 f., 63, 69, 81, 85, 90, 93, 101–105, 116–119, 125, 130–135, 140–146, 148–150, 153–156, 159–163, 168, 171, 177, 181, 183, 206, 209–211, 214, 218–222, 225, 227
Adequacy conditions   6, 8, 10–13, 17–19, 23, 28, 30, 36, 40, 52, 55, 58, 60, 93, 100, 103, 112 f., 140, 150, 198, 225
Agentive abilities   6, 10, 13 f., 28, 58–61, 80, 92–95, 97 f., 100, 102, 109, 112, 118, 140, 142, 144–147, 150, 152–162, 166, 170–172, 177 f., 181, 192, 194 f., 202, 206, 211, 221, 225–227
Alternate Possibilities Contextualism   15, 218, 221, 223
Analysis   2, 6–10, 14, 16, 31, 39 f., 47, 50, 52, 55, 58, 61, 73, 77, 101 f., 138, 141, 154 f., 159, 164, 171, 176, 181–184, 186, 188, 199, 209, 221 f.

Brainwash   50–52, 111, 170, 226

Categorical properties   8 f., 63
Coma   49–52, 55, 78 f., 89, 108–112, 126, 151, 170 f., 194–197, 209, 226
Compatibilism   3, 38, 163, 167 f., 180, 211, 214 f., 218–221
Complexity   64, 95, 134, 194, 196, 213
Compossibility   12, 66, 218
Compossibility   66, 68, 216–218
Conative abilities   117, 120–122, 126, 151
Concepts   1, 9, 18, 121, 183, 214, 218
Conditional analysis   3–5, 11–14, 30 f., 33, 37–40, 43–47, 49–63, 71 f., 74, 77–80, 82, 89, 93 f., 97–100, 103, 108–113, 135, 140, 150, 154, 163, 176, 180–182, 184, 188–192, 194–200, 212 f., 225
Context sensitivity   4, 10, 12 f., 23, 26–28, 33 f., 36, 40, 44, 55 f., 61, 65–73, 81 f., 87, 90, 93, 97–100, 103, 105–109, 112, 117, 120 f., 126, 131, 150, 157 f., 162, 169 f., 172, 178, 180 f., 186, 188, 190, 192–194, 196 f., 200 f., 215–223, 225
Contextualism   15, 69 f., 98, 215 f., 218 f., 221–224
Control   29, 34, 145, 148 f., 193, 204, 210, 219 f.
Counterfactual   1, 3 f., 9, 11 f., 30–33, 37–47, 49–51, 53–55, 57–63, 70, 74, 78, 95, 99 f., 108, 110–112, 135, 162 f., 176, 179–181, 188–192, 194–200, 212

De dicto   13, 149, 153, 162, 221, 225
Degrees   10, 12 f., 23–29, 33, 36, 40, 44, 55–57, 61, 67, 71 f., 80–84, 90, 93, 99, 103–108, 112, 126–128, 134 f., 150, 152, 157 f., 162, 170, 178, 180, 187 f., 190, 194, 200, 202–205, 207 f., 215, 225–227
Dependency   134
De re   13, 149, 153, 162, 221, 225
Determinacy   134 f.
Determinism   3 f., 15, 38, 117, 167 f., 180, 209–211, 213–215, 218–220, 222 f.
Deviant causal chains   101, 150
Disagreement   18, 51, 75 f.
Dispositions   2 f., 8 f., 11, 14, 23, 32–34, 37, 46–48, 59, 71, 84, 94, 103 f., 106, 141, 145, 162–170, 174, 176, 178–185, 187–192, 194, 199–202, 212 f., 225

Empty sets   27, 71, 109–111, 170 f.
Exercise   20, 29–31, 34, 44 f., 48, 53 f., 58 f., 73–75, 79, 89, 90, 94, 116, 121, 124, 135 f., 140 f., 143, 148, 152, 154 f., 159, 170, 176, 186, 205, 220
Existential requirement   108, 110 f., 118, 151, 164, 170–173, 178 f., 226 f.

Explanatory challenges 11, 14, 17, 30, 32, 34, 36, 162, 176, 179
Extensional adequacy 10 f., 17 f., 36
Extrinsicality 21, 98, 112, 115 f., 132, 151, 159, 167 f., 179, 189

Failure 1, 7, 10, 11 f., 14, 16, 23, 40, 44, 45 f., 53 f., 60, 49–56, 58–61, 63, 72, 74–76, 78, 81 f., 84 f., 87, 90, 109–111, 114 f., 117, f., 122, 124–126, 137–139, 142 f., 146 f., 150, 164 f., 167, 172 f., 191, 195–197, 203, 207, 211, 220, 226
Finks 140 f.
Fluke 22, 117
Freedom contextualism 219, 221, 223 f.
Free Will 2–5, 15, 38, 63, 69, 163, 167 f., 180, 185, 210, 214 f., 218–224
Functions 85 f., 130, 139, 157, 170, 198

General abilities 19–24, 35, 52–55, 61, 72–74, 76 f., 84, 87–90, 99, 112–117, 119–126, 128, 136, 139, 151 f., 158 f., 166 f., 171–174, 202, 206 f., 210 f.
Grandfather paradox 15, 214–218

Hypothetical circumstances 128–135, 152, 207

Ignorance 146–149, 215
Impeded intentions 44, 49, 51, 55, 61, 72, 77–79, 89, 93, 108 f., 110–112, 135, 151, 194–197, 212 f., 225 f.
Incompatibilism 3, 15, 214 f., 218–220, 222 f.
Infinite sets 173
Intentions 6, 10 f., 13 f., 28–30, 33, 39 f., 44–46, 49–54, 58–61, 72, 74, 78–81, 89 f., 92–106, 108–113, 117–120, 122, 124, 126, 132, 136, 140–163, 166, 168, 170–172, 174 f., 177 f., 187, 190 f., 194 f., 201–203, 205, 207, 209, 213, 221, 223, 225–228
Intrinsicality 21, 71, 73, 98, 112 f., 115 f., 122, 126, 132, 151, 159, 164, 166–168, 173, 176, 179, 189, 192–194, 196, 200 f., 203

In view of 4, 12 f., 34, 39, 47, 53, 65–67, 69, 71, 74, 78–81, 89, 93, 98, 108, 112 f., 115–119, 128–133, 136, 150–152, 162, 167 f., 172 f., 179, 192, 202–204, 207, 217, 222 f.

Limit assumption 42 f., 45, 136
Locality 134–136

Masks 40, 44–49, 53–57, 60 f., 71–77, 79, 89, 93, 124 f, 135 f., 141, 152, 188, 190, 194, 196, 198, 225
Mere behavior 28 f., 141, 146, 155 f., 159–161, 163, 168, 177
Metaphysics 8 f., 205
Modal auxiliaries 2, 5, 34, 63
Modal base 65–67, 69–71, 73–80, 89, 98, 113 f., 116 f., 119–121, 124, 126, 128 f., 132, 222 f.
Modal force 64, 68, 77, 99
Modality 1 f., 5 f., 8 f., 11–14, 22 f., 31–33, 39–41, 60, 63–67, 69, 72, 75–77, 85 f., 92–100, 102–104, 108–110, 117, 137 f., 140, 143 f., 150, 154 f., 159 f., 162 f., 172, 175, 177, 187, 199, 201, 205–208, 211–213, 217, 225
Motivation 2 f., 11 f., 30 f., 33, 38 f., 50 f., 58, 60, 93, 100–103, 111, 119–121, 126, 150 f., 191, 210, 220 f., 225

Necessary condition 6, 46, 49, 52, 54 f., 60 f., 171, 227
Necessity 5, 32, 64, 87
New dispositionalism 3, 14, 32, 163, 167 f., 180–182, 184 f., 192
Non-agentive abilities 6, 10, 12, 14, 28–30, 36, 40, 44, 58–61, 72, 77–79, 81, 90, 92 f., 141, 154–164, 166, 171 f., 177 f., 181, 191, 211, 213, 225, 227
Normativity 93, 168

Obsession 203, 205
Opportunity 20 f., 87, 115, 119–122, 193 f., 196, 214
Options 22, 24 f., 59, 185, 205–213
Ordering source 67 f., 71, 83 f., 90, 174

Particular abilities   117 f., 120, 122 – 125, 151
Possibilism   4 f., 12 – 14, 37, 63, 66, 68 – 91, 93, 97 – 100, 112 – 115, 137 f., 140, 150, 152, 180, 182, 198, 200, 214, 216 f., 225 – 227
Possibility   1, 4 f., 9, 12, 31 f., 35 f., 42, 63 – 70, 72, 77, 80 f., 83 – 88, 90 f., 99, 137 f., 140, 152, 175, 177, 200, 216 f.
Possible worlds   8, 41, 44, 50, 64 – 71, 74 f., 77 – 79, 83 – 86, 89, 96 – 98, 107, 114, 138, 175, 187, 199 f., 216 f.
Power   4, 16, 38, 115, 119, 166, 206
Proportions   6, 13 f., 94 – 100, 102 – 114, 117, 119 f., 122, 124, 126, 130 – 132, 136 – 161, 163 – 165, 167 f., 171 – 175, 177 f., 182, 185 – 205, 207 f., 212 f., 217 f., 221, 223, 225 – 228

Reliability   7, 22, 25 f., 56 f., 81 – 84, 90, 103 – 108, 114, 123 f., 158, 183 f., 217
Restrictions   1, 4, 12, 35 f., 63 – 65, 67 f., 70 f, 72, 78, 80 f., 85, 89 f., 97 f., 110, 118, 132 f., 137 f., 140, 155, 160, 177, 187, 200 f., 207, 216 f., 225, 227

Scales   26 – 28, 55 f., 61, 67, 74 – 76, 82 f., 90, 103, 108, 126, 174, 217
Semantics   2, 5, 7, 10 f., 31, 35 – 37, 40 – 44, 49, 53, 63 – 66, 69 f., 83, 85, 176 f., 179, 195, 199, 219, 224

Situations   2, 13 f., 21 f., 25 f., 53, 84, 95 – 100, 102 – 114, 116 – 120, 122 – 147, 149 – 165, 167 f., 170 – 175, 177 – 179, 182, 185 – 187, 189, 192 f., 195, 198 – 208, 211 – 213, 217, 221 – 223, 225 – 228
Solidity   6, 17, 71, 122 f.
Specific abilities   10, 12 f., 19 – 25, 35 f., 40, 44, 52 – 55, 61, 72 – 77, 86 f., 89 f., 93, 98, 112 – 128, 133 f., 135 f., 151 f., 157 – 159, 167 f., 171 f., 192, 196, 201, 206, 225
Specifier approach   47 – 49, 61
Stimulus   29, 33, 46 f., 94, 104, 141, 156, 162, 164, 189 – 191, 199 f.
Success   6, 10 – 15, 18, 22, 26, 46, 49 f., 56, 77, 87, 90, 92 – 96, 99 f., 102 – 104, 106 – 111, 114 f., 117, 120, 122 f., 124 – 126, 130 – 133, 135 – 138, 140 – 147, 150 – 166, 168 – 192, 194 – 205, 207 f., 211 – 215, 217 f., 221 – 227
Sufficient condition   6, 49, 52, 61, 226

Trigger   14, 47, 70, 92, 154 – 158, 160 – 166, 168 f., 171 – 173, 177 – 179, 182, 186, 188, 191 – 194, 197, 201, 205, 212, 225, 227 f.
Truth conditions   1 f., 6 f., 9 f., 22 f., 27, 41 – 43, 45, 63 f., 71, 111, 135, 138 – 140, 154, 218 f., 222 f.

Virtue epistemology   14

www.ingramcontent.com/pod-product-compliance
Lightning Source LLC
Chambersburg PA
CBHW030540230426
43665CB00010B/965